AN INTRODUCTION TO FISHES

By

H.S. Bhamrah
Kavita Juneja

Edited by

Balvinder K.

ANMOL PUBLICATIONS PVT. LTD.
NEW DELHI - 110 002 (INDIA)

ANMOL PUBLICATIONS PVT. LTD.
4374/4B, Ansari Road, Daryaganj
New Delhi - 110 002
Ph.: 23261597, 23278000
Visit us at: www.anmolpublications.com

An Introduction to Fishes

First Published, 1990
Reprint 1991, 1992, 1994
Second Revised Edition, 2001
Reprint 2004

PRINTED IN INDIA

Published by J.L. Kumar for Anmol Publications Pvt. Ltd., New Delhi - 110 002 and Printed at Mehra Offset Press, Delhi.

Contents

Preface

We feel pleasure in bringing out the second revised edition of the present book on the instance of our learned teachers. Besides rectifying the errors of first edition, new material is added here and there in the light of recent development and researches in the field.

This text is designed to approach the morphology, anatomy, physiology and development of the *Fishes* in a coherent way since it is most natural for a student in studying the architectural elements of an animal body in the quest to know how they function. This text intends, therefore, to enable the student, through his own natural appetency, to correlate form and function so that he possesses for subsequent work in either anatomy or physiology a firm foundation for a real edifice of knowledge.

The book features both a text and a laboratory guide. It deals with the structures and physiological phenomena of each system. Origin, adaptative radiation and other general topics are also dealt in details. It is hoped that this book will not only meet the requirement of Indian students but will also be useful as a guideline to the teachers in their teaching.

There can be no claim to originality except in the manner of treatment and much of the information has been obtained from the books and scientific journals available in different libraries.

We express our thanks to our friends and colleagues whose constant inspiration have initiated them to bring out this book.

The authors are also grateful to Dr. J. Bahadur, Professor and Head, School of Studies in Zoology, Jiwaji University Gwalior, Dr. A.K. Pandey, Professor & Head, Vikram University, Ujjain for their valuable suggestions.

All the constructive criticisms and suggestions for improvement from learned teachers and scholars are welcome.

H.S. Bhamrah
Kavita Juneja

1

Agnatha

The Agnatha are earliest known vertebrates and are characterized by the absence of jaws. They are divided into several classes of which only Cyclostomata is represented by living examples while remaining classes have extinct forms. Thus the fossil Agnatha are collectively known as *Ostracoderms*.

CLASS-OSTRACODERMI

They are the earliest known vertebrates who flourished from Ordovician to Devonian Periods and became extinct about 300 million years ago.

General Characters

1. The shape of body was fish-like in general appearance but dorso-ventrally flattened.
2. The paired fins were absent. Median dorsal and caudal fins were present. Tail was *heterocercal*.
3. They had bony dermal plates on skin.
4. The skull was made up of bones and cartilages. Vertebral column and girdles were absent.
5. There was a slit-like mouth at the extreme front end. The jaws were absent.
6. The gill-slits were 10 pairs and all had similar gill-pouches.
7. They had single nostril on the top of head.
8. 10 pairs of cranial nerves were present.
9. Paired eyes were present on the top of the head. Median pineal eye was also present.
10. Ears with two semi-circular canals were present.

Classification

Class-Ostracodermi is divided into two sub-classes:

Sub-class-1. Monorhina

A single, large, slit-like median nostril was present on the top of the head.

This sub-class is divided into 3 orders :

Order. 1. Euphanerida

1. About 18 cm. long, fusiform, naked body.
2. Notochord was present.
3. Two lateral and a median dorsal fin fold were present.

   Example : *Jamoytius.*

Order. 2. Anaspida

1. Small body measuring 15 ms. in length.
2. Head naked or covered with plates.
3. Tail *hypocercal.*

   Example: *Birkenia.*

Order. 3. Osteostraci or Cephalaspida

1. A single piece head shield.
2. Tail heterocercal.
3. Several pairs of gill-slits were present.

   Examples: *Cephalaspis, Tremataspis.*

Sub-class-2. Diplorhina

Two separate nasal openings were present.

Order. 1. Heterostraci

1. Head encased in a shield.
2. Tail *hypocercal.*
3. Eyes widely spaced.
4. One pair of gill-slits only.
5. Sensory organs were absent.

   Example : *Pteraspis.*

Order. 2.Coelolepida

1. Skin with several minute placoid scales.
2. Tail *hypocercal* or *heterocercal.*

   Example : *Phlebolepis.*

Affinities

The ostracoderms lived from Silurian to Devonian periods. Possibly they arose from unarmoured ancestors like *Jamoytius* during the Ordovician

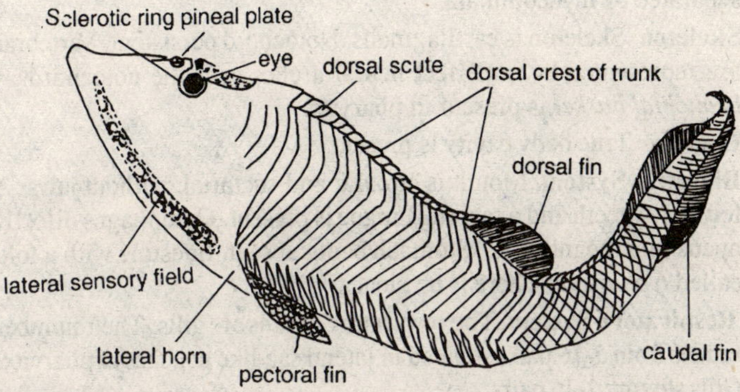

Fig. 1.1. *Cephalaspis.*

and quickly become adapted to various modes of life. During Devonian period, gnathostomes were also evolved simultaneously and the ostracoderms could not compete with them so that they became extinct.

The ostracoderms resembled the cyclostomes in many characters like single-median naris, pineal eye, mouth without jaws and unpaired fins, Thus they appear to be closely related to living Agnatha (i.e. cyclostomes) that is why cyclostomes and ostracoderms are placed together in the super-class Agnatha. Cyclostomes are, therefore, considered as degenerate descendants of some Cephalaspid or Anaspid-like ancestors.

CLASS-CYCLOSTOMATA

Class-Cyclostomata or Marsipobrachii are the living representatives of super class-Agnatha. These are characterised by the presence of suctorial mouth (without functional jaws). They are the most primitive of all living vertebrates.

General Characters

Class Cyclostomata includes lampreys and hagfishes. These animals possess the following characters:

1. **Body shape.** The body is long, slender, eel-like with head and trunk nearly cylindrical and tail laterally compressed. Neck is lacking.
2. **Skin.** The skin is smooth, soft, slimy and without scales. Unicellular mucous glands are present.
3. **Fins.** Paired fins are absent. Median fins are present supported by cartilaginous fin rays.

4. **Muscles.** Trunk and tail muscles are arranged in 'E' shaped myotomes separated by myocommata.
5. **Skeleton**. Skeleton is cartilaginous. Notochord persistent. Vertebrae are represented by imperfect neural arches over the notochord. A *branchial basket* is present in pharynx.
6. **Coelom** . True body cavity is present.
7. **Digestive System.** Mouth is circular and suctorial, without jaws. A few horny teeth and a rasping tongue is present. Oesophagus directly opens into the intestine. Stomach being absent. Intestine with a fold called *typhlosole*. There is no cloaca.
8. **Respiratory System.** The respiratory organs are gills. Their number ranges from 5-16 pairs situated in lateral sac-like pouches of pharynx. Gills slits are 1-16 pairs.
9. **Circulatory System.** Heart is two chambered,with an auricle and a ventricle. Conus arteriosus is absent, but *sinus venosus* is present. Many aortic arches are present in gill region. White blood corpuscles (leucocytes) and nucleated circular red blood corpuscles (erythrocytes) are present in blood.
10. **Body Temperature**. Poikilothermous i.e. cold blooded animals. Their body-temperature is variable.
11. **Nervous System.** Brain is differentiated with a very small cerebellum. 8-10 pairs of cranial or cerebral nerves.
12. **Sense Organs**. Single median olfactory sac with a single median nostril. Eyes may be functional or vestigial and sunk beneath the skin. Ears may have one or two semicircular canals. Lateral line system is present.
13. **Excretory System.** Sexes may be separate (unisexual) or united (hermaphrodite). Gonad (testis or ovary) is single, large without gonoduct. Fertilization external.
14. **Development**. Direct or indirect with a larval stage.

Classification

Class-Cyclostomata is divided into two sub-classes :

(1) Sub-class Petromyzontia and (2) Sub-class Myxinoidea.

Sub-class-1. Petromyzontia

1. Mouth is ventral, surrounded by a large suctorial buccal funnel. The inner surface of funnel is beset with many conical, yellowish, horny teeth.
2. Nostril and the nasal sac are dorsal. The nasal sac does not open behind into the pharynx.

3. There are seven gill-slits on each side behind the eyes. They open into a respiratory tube lying ventral to the oesophagus.
4. They possess one or two well developed dorsal fins.
5. Branchial basket is complete.
6. Dorsal and ventral roots of spinal nerves are separate.
7. Eyes are functional. Pineal eye is well developed. Ear has two semicircular canals. Lateral line system is present.
8. Sexes are separate. Eggs are small and numerous. Segmentation is holoblastic. Development is indirect through a larval stage.
9. Both marine and fresh water forms are parasitic in habit.

   Examples :*Petromyzon marinus, Lampetra fluviatilis, Caspiomyzon, Mordiacia* etc.

Sub-class. 2-Myxinoidea

1. The mouth is terminal without buccal funnel and has four pairs of tentacles around it.
2. Nostril terminal. Nasal sac opens behind into pharynx.
3. There are 1-15 gill apertures on each side, opening directly into pharynx.
4. Eyes are vestigial and sink beneath the skin.
5. Dorsal fin is absent or only feebly developed.
6. Branchial basket is reduced.
7. Brain is primitive as cerebrum and cerebellum being unrecognizable. Dorsal and ventral roots of spinal nerves are united. The lateral-line system is reduced.
8. Ear with one semicircular canal.
9. Large shelled eggs. Segmentation is *meroblastic*. Development without any larval stage i.e. direct.
10. They are marine.

    Examples: *Myxine glutinosa, Bdellostoma* or *Eptatretus cirrhatum, Paramyxine.*

MYXINE

Phylum	-	Chordata
Group	-	Vertebrata
Super class	-	Agnatha
Class	-	Cyclostomata
Sub-class	-	Myxinoidea
Genus	-	*Myxine*

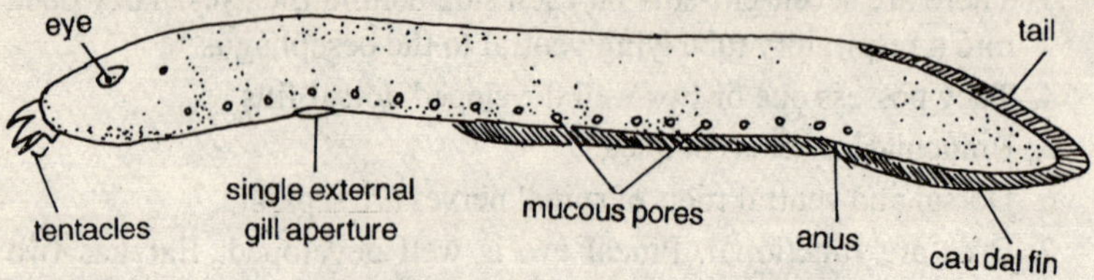

Fig. 1.2. *Myxine*

1. It is commonly known as '*hagfish*'.
2. They are marine found burried in the muddy bottom of sea.
3. Body is cylindrical,eel-like, soft, without any scale and measuring about 60 mm in length.
4. Mouth is sub-terminal without buccal funnel and has four pairs of *tentacles* around it. Jaws are absent but a single hooked palatine tooth is present.
5. Eyes are degenerate and covered with a thick skin due to dark and bottom dwelling habit.
6. A single median nostril lies close to the mouth.
7. Six pairs of gill-pouches are present behind the head region but open to the exterior by one pair of external gill-slits.
8. Mucous glands are mumerous which open all along the sides of body length. They secrete mucous or slime hence another name is given as '*slime-eel*'.
9. A feebly developed median ventral fin and a caudal fin are present. Fin-rays are absent.
10. They are scavanger. They attack injured or dead fishes and burrow into their body for eating of viscera and muscles. Thus they are *parasitic* or *quasiparasitic* and generally remain attached to the body of other fishes.
11. They are bisexual or hermaphrodite with a single *ovotestis* without gonoduct. They are *protandrous* as sperms mature earlier. The eggs are large enclosed in shell, bearing hooks by which they attach themselves to weeds.
12. Development is direct i.e. without a larval stage.
13. They are found along the ssea-coasts of N. Atlantic, American and Japanese sea waters.

BDELLOSTOMA

Phylum	-	Chordata
Group	-	Vertebrata
Super class	-	Agnatha
Class	-	Cyclostomata
Sub-class	-	Myxinoidea
Genus	-	*Bdellostoma* or *Eptatretus*

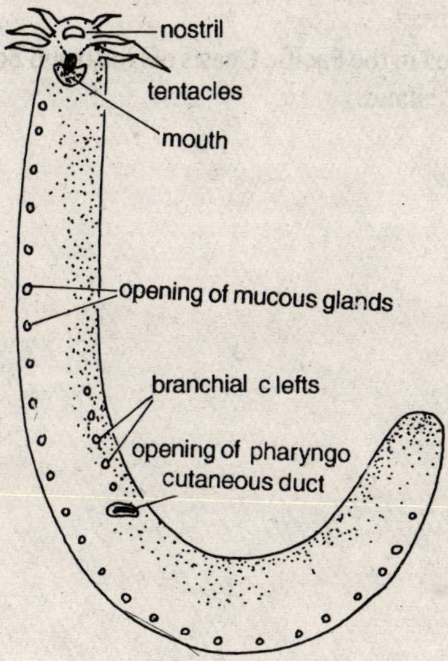

Fig. 1.3. *Bdellostoma*.

1. It is also commonly known as '*hagfish*'.
2. Body is elongated, soft, eel-like, without scales and measuring upto one metre in length.
3. The tail is laterally compressed.
4. Mouth is sub-terminal, without jaws but a number of teeth are present. It is surrounded by four pairs of tentacles.
5. Eyes are vestigial and covered with the skin.
6. Single nostril which opens into nasal sac. The nasal sac opens into the pharyrnx.
7. 6-14 pairs of gill-pouches are present which open out side by separate gill-slits.

8. Behind the last external gill-slit of left side there is an opening of pharyngocutaneous duct which opens into pharynx.
9. Entire body contains double rows of mucous glands which secrete mucous.
10. They are nocturnal scavenger, feed on dead or dying fishes by sucking their viscera.
11. Hermaphrodite with single ovotestis. They are protandrous.
12. The eggs are large enclosed in shell. Cleavage is meroblastic. Development is direct.
13. They are distributed in the Pacific Coasts of North and South America, Africa and New Zealand.

2

Petromyzon

SYSTEMATIC POSITION

Phylum	-	Chordata
Group	-	Vertebrata
Super class	-	Agnatha
Class	-	Cyclostomata
Sub-class	-	Petromyzontia
Genus	-	*Petromyzon*

Distribution

These are commonly known as 'Lampreys', 'Lamperns', 'Sandprides', 'Lamper-eels' etc. These are most primitive living vertebrates. Three species of lampreys are common in European river system. *P. marinus* (marine), *Lampetra fluviatilis* (fresh water) and *L. planeri* (lesser fresh-water).

Habits and Habitat

Petromyzon is a predator and feeds on the blood of fishes i.e. *saguivorous* in habit. It is often found attached to the host when the latter is caught. It swims actively in the water by the lateral undulations of its highly muscular body. It breeds only once in its whole life. During breeding season the marine *Petromyzon* migrates to the fresh water for reproduction and laying of eggs.

EXTERNAL FEATURES

The body is elongated, cylindrical and eel-like, with a laterally compressed tail. The length of the animal varies from 45 cm as in *L. planeri* to a metre in *P. marinus*. The skin is naked i.e. without any scale. The skin is slimy and contains unicellular glands which secrete mucous. The colour of the body on dorsal side is greenish, brown with patches of dark colour. The colour on ventral side is pale or white.

The body in differentiated into three parts : the head, trunk and tail.

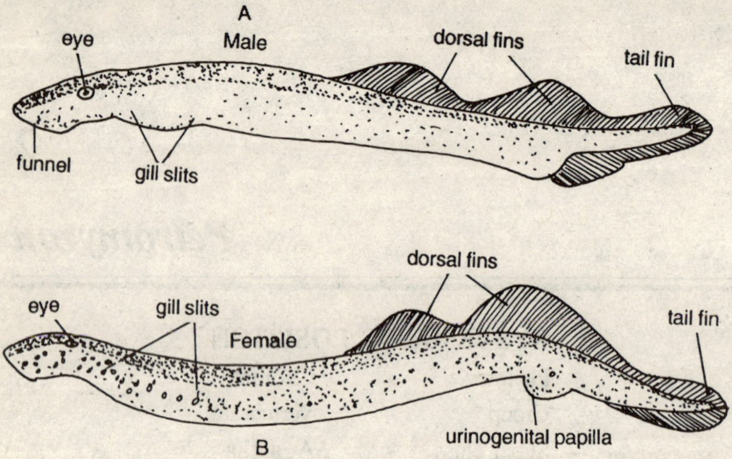

Fig. 2.1. *Petromyzon.* External features A—Male, B—Female.

Head

The head bears a mouth, a single median naris, a pair of eyes and seven pairs of gill-slits.

Mouth : The mouth is situated on the ventral side of the anterior end, in the centre of a basin-like depression called the *buccal funnel.* The buccal funnel is surrounded by a marginal membrane outside of which

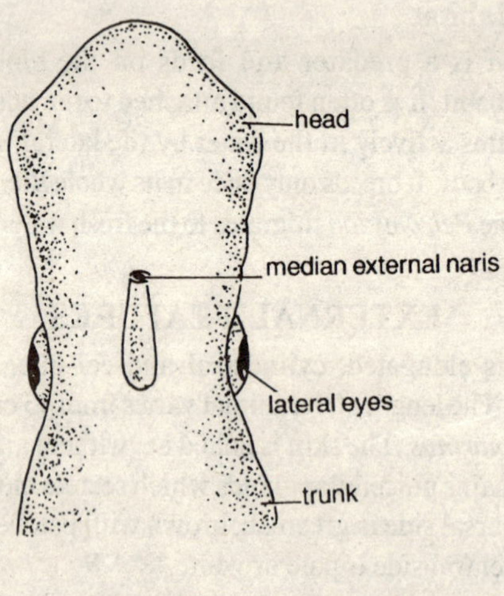

Fig. 2.2. *Petromyzon.* Dorsal view of the head.

arise smaller processes called the *oral fimbriae* and some longer, sensory processes known as *cirri*. The jaws are absent but radiating rows of yellow horny teeth are present on the inner surface of buccal funnel. These teeth are present on the cartilaginous pads and are principally mesodermal in origin. From the bottom of the funnel there projects a prominence, called the *tongue*. It also bears larger teeth and having immediately above it the narrow mouth.

Nostril or Naris : On the dorsal surface of the head is a single median nostril. It opens behind into a *nasal sac* present in skull which does not open into the pharynx i.e. it remains closed behind. Immediately behind the nostril there is a transparent area of the skin which indicates the position of the pineal organ.

Eyes : The eyes are situated on the lateral sides of the head. They have no eye-lids, but are covered by transparent skin.

Gill-slits : There are seven pairs of gill-slits present on either side of head just behind the eyes. They open into the gill-pouches.

Trunk

The trunk lies behind the head and in front of tail region. There is no traces of paired appendages as paired fins are present in fishes. Two dorsal fins of roughly equal size (separated by a notch) are present. These

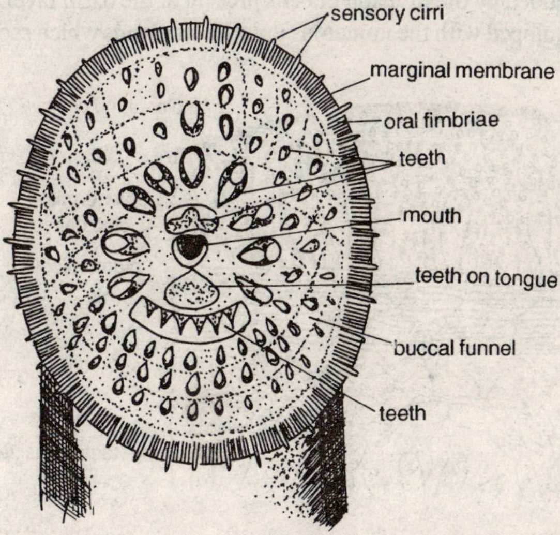

Fig. 2.3. *Petromyzon*. Buccal funnel.

are known as *Ist dorsal* and *2nd dorsal fin* respectively. Around the tail there is a *caudal fin*. The second dorsal fin being continuous with the caudal. All these fins are median and unpaired and are supported by cartilaginous *fin-rays*. On the ventral side at the junction of trunk and tail there lies a small aperture known as *anus*. Behind the anus is present a *urino-genital papilla* on which the urino-genital aperture is situated.

Tail

The laterally compressed part behind the anus is called the tail. It bears a *caudal* or *tail fin*. It is continuous with the 2nd dorsal fin.

The segmental sense organs take the form of *lateral line system.* these are not enclosed in a canal (as in fishes), but are exposed to exterior in the form of a row of lateral pits extend along each side of head, trunk and tail.

The sexual dimorphism is seen during breeding season. In the female an *anal fin* is developed behind the urinogenital aperture. The margins of anus become reddish. In the male a *penial tube* is formed and the base of dorsal fin is thickened.

BODY-WALL

The skin is soft and slimy lacking in exoskeletal structures like scales etc. It consists of *epidermis* and *dermis*. The epidermis is several layers in thickness. Keratin does not occur in the epidermis so that all the cells of epidermis are alive. The outer layer of the skin is shed from time to time, new cells are added by the division of cells present at the basal layer. The epidermis is equipped with the numerous unicellular glands which secretes

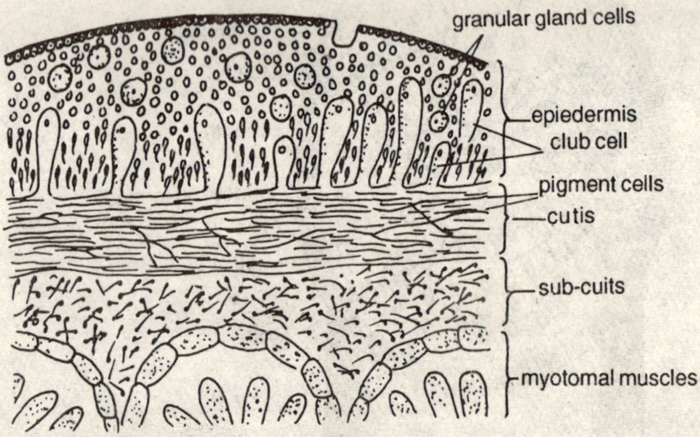

Fig. 2.4. *Petromyzon.* V.S. Skin.

mucous which makes body slimy. In addition two more types of secretory cells are present, one is the *granular cells* of unknown function and the other is *club-cells*, which probably help in sliminess, having cytoplasm.

Underlying the epidermis is a layer of collagen and elastic fibres which forms the *cutis*. This layer also contains *chromatophores*. The chromatophores are star-shaped, bearing pigments in them. The pigment changes its position according to the intensity of the light. Thus the colour of lamprey changes from pale to dark or *vice versa*.

The dermis consists of connective tissue, muscle fibers and blood vessels.

ENDOSKELETON

The entire skeleton in *Petromyzon* is made up of cartilage.

Skull

The skull is very primitive in structure and consists of cartilages and membranes. The floor of cranium is formed by a *basal plate* which is made

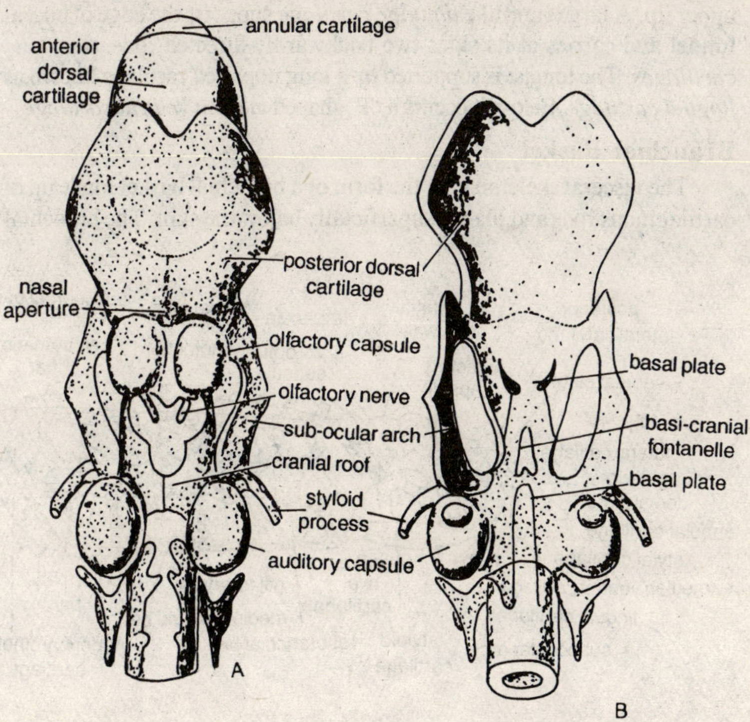

Fig. 2.5. *Petromyzon*. Skull A—Dorsal view, B—Ventral view.

up by the fusion of parachordals and trabeculae. An aperture called *basicranial fontanellae* is present in the middle of the basal plate. The nasohypophysial sac traverse through this aperture. Sides of the basal plate extend upwards to form the lateral walls of cranium. The roof of the cranium is covered all over by membrane except at one place. A narrow bar of cartilage lies across the hinder part of the membranous roof. A pair of *optic capsules* are joined with the posterior end of the basal plate. A median *olfactory capsule* is attached to the cranium. It is made up of the convexo-concave plate, and is pierced by paired apertures for olfactory nerves.

There is a *sub-ocular* arch which extending outwards and downwards from each side of'the basal plate to support the eyes. From its posterior end a slender *styloid process* passes directly downwards and is connected at its lower end with a small *cornual cartilage*. A raised plate - like structure called *posterior lateral cartilage* is found above the sub-ocular arch.

Connected with the anterior end of the basal plate is the large bilobed *posterior dorsal plate*. Below and projecting in front of basal plate is *anterior dorsal plate*. A pair of *anterior lateral cartilage* support the upper lip. A large ring-like *annular cartilage* supports the edge of buccal funnel and carries at its sides two backwardly directed rods, the *stylet cartilages*. The tongue is supported by a long unpaired cartilage known as *lingual cartilage*. Below it occurs a 'T' shaped *median ventral cartilage*.

Branchial Basket

The visceral skeleton is in the form of a branchial basket made up of cartilaginous rods and placed superficially below the skin. The branchial

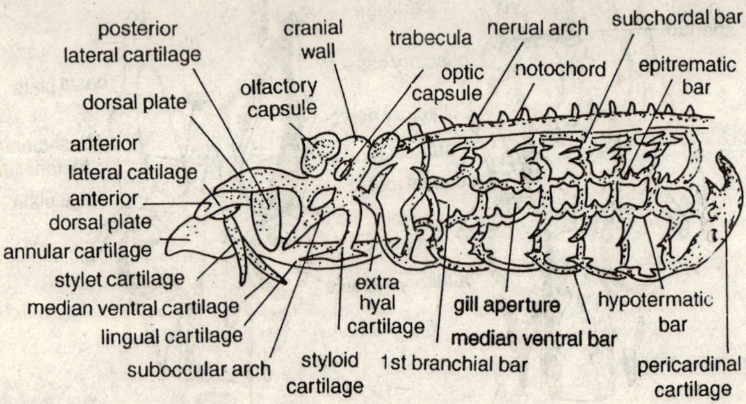

Fig. 2.6. *Petromyzon*. Branchial Basket.

basket is composed of nine vertical bars of cartilage that are curved irregularly and are connected with each other by means of four longitudinal bars. The first vertical bar lies just behind the styloid cartilage. The second is infront of the first gill-slits. The last vertical is connected with a cup-like pericardial cartilage, which supports the posterior and lateral walls of the pericardium. The longitudinal bars are : *subchordal bar* lies alongside the notochord and is connected in front with the cranium; the *epitrematic bar* and *hypotrematic bars* are placed above and below the gill-clefts respectively, and *median ventral bar* situated close to the middle ventral line and is partly fused with its fellow of opposite side.

The branchial basket supports the outer part of branchial passage.

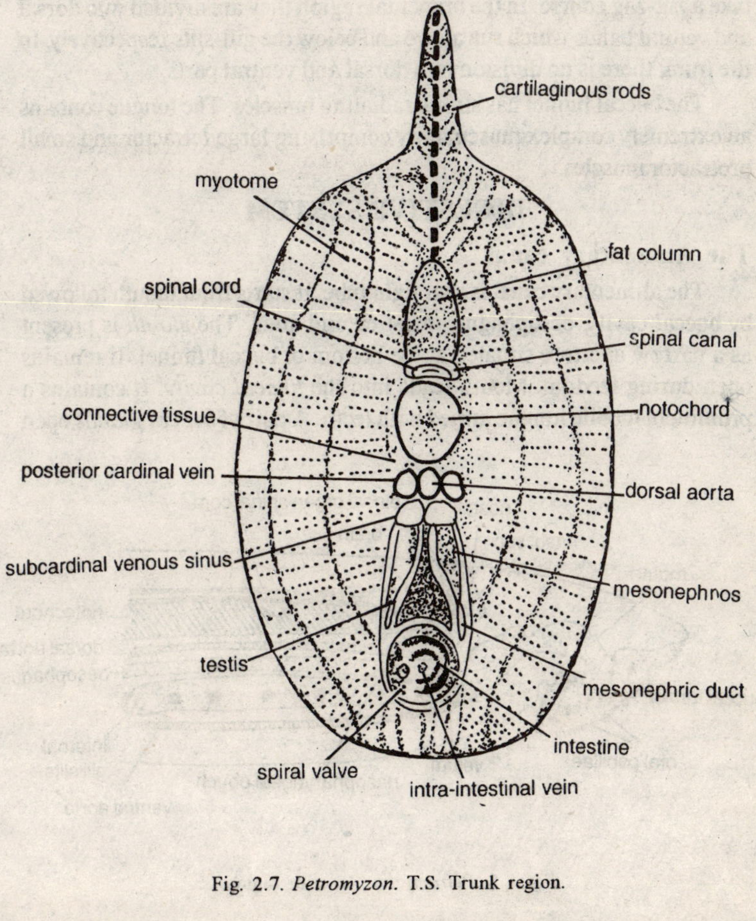

Fig. 2.7. *Petromyzon*. T.S. Trunk region.

Vertebral Column

The axial skeleton is simple. There is a persistent *notochord*. It is enclosed by a tough sheath composed of inner fibrous and outer elastic layer. Attached to the sides of notochord are small, vertical rods of segmentally arranged cartilage that bound the spinal canal. In the caudal region these fuse into a single plate perforated by foramina for the exit of spinal canal. For the rest of its length, the spinal canal is enclosed only by tough, pigmented connective tissue.

The fins are supported by slender rods of cartilage called *fin-rays*.

Musculature

The muscles of the trunk and tail are arranged in myomeres which take a zig-zag course. In the branchial region they are divided into dorsal and ventral bands which run above and below the gill-slits respectively. In the trunk there is no division into dorsal and ventral parts.

The buccal funnel has strong radiating muscles. The tongue contains an extremely complex musculatory comprising large retractor and small protractor muscles.

DIGESTIVE SYSTEM

The Alimentary canal

The alimentary canal is a straight tube. It starts from mouth followed by buccal cavity, oesophagus, intestine and anus. The *mouth* is present as a narrow aperture situated at the bottom of buccal funnel. It remains open during feeding. Mouth leads into the *buccal cavity*. It contains a prominent mobile tongue armed with teeth. A pair of *buccal glands* open

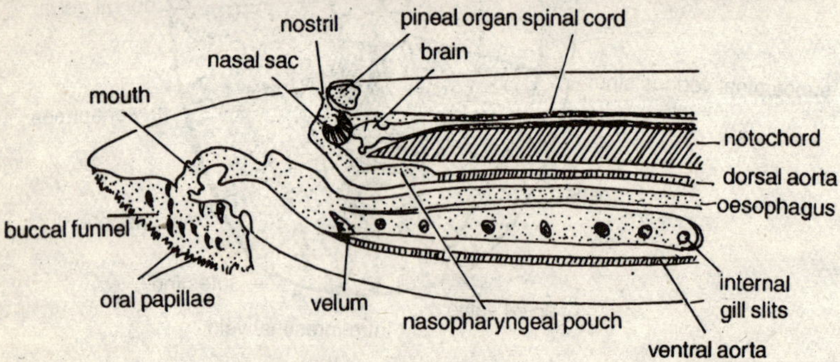

Fig. 2.8. *Petromyzon*. Anterior region.

below the tongue which secrete a substance that prevents the coagulation of blood. Posteriorly the *buccal cavity* opens into two tubes, the dorsal of these is *oesophagus* and the ventral is the *pharynx* or *respiratory tube* guarded by a velum. The oesophagus is followed by the *intestine* through a valvular aperture. The intestine is straight tube. Its anterior end is slightly dilated. A true stomach is absent. A fold called the *typhlosole* projects into the lumen of intestine. The typhlosole follows a slightly spiral course along the intestine, therefore, it is also called *spiral valve*. The intestine opens out by a small aperture called *anus*.

Digestive glands

Besides *buccal glands*, a large bilobed *liver* is also present. Gall-bladder and bile duct occur in larva, but are at least sometimes absent in adult. There is no separate *pancreas* but the *zymogen cells* and *follicles of Langerhans* are embedded in the gut epithelium. Zymogen cells secrete proteolytic enzymes while the follicles of Langerhans control carbohydrate metabolism.

The typhlosole contains reticular tissue in which blood cells are formed. Thus this tissue represent spleen of *Petromyzon*

RESPIRATORY SYSTEM

The respiration is done by seven pairs of *gill-chambers* or *gill-pouches*. Each pouch opens to the exterior by an aperture called *external gill-slit*. Internally, the gill- pouches open into the pharynx or respiratory tube by *internal gill-slits*. The pharynx is closed at the posterior end. In the larval stage the pharynx remains open behind into the oesophagus. The passage between the pharynx and buccal cavity is guarded by *velum*

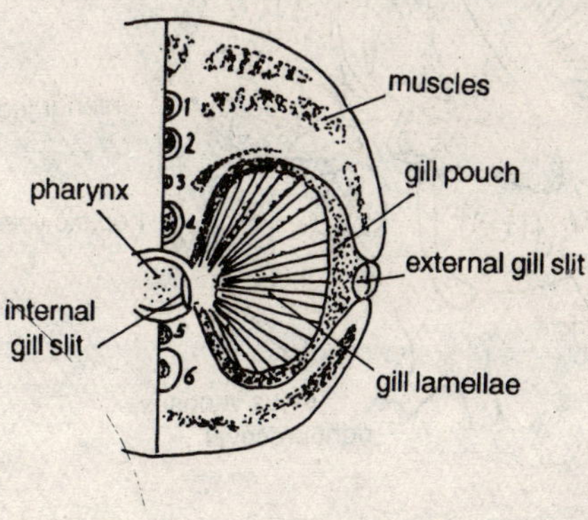

Fig. 2.9. *Petromyzon*. Gill pouch.

to check the entry of food into pharynx. The gill-pouches are separated from each other by *inter-branchial septa*. Each chamber bears on its inner surface a number of *gill-lamellae*.

The respiratory water current enters and leaves the gill-pouches through external gill-slits. It is due to contraction and relaxation of gill-pouches alternately. Water washes the gill-lamellae and this brings about oxygenation of blood passing through them. The cartilaginous branchial basket also plays an important role in respiration.

CIRCULATORY SYSTEM

The circulatory system is composed of a pumping organ, the *heart*, few *blood vessels* and *blood*. The heart is enclosed in a thick-walled *pericardial sac*. As the heart is longer than the pericardial sac, it is slightly bent on itself to form the 'S' shaped structure. The posterior and lateral walls of pericardial sac are supported by *pericardial cartilage*. The heart is made up of *atrium*, *ventricle* and *sinus venosus*. The atrium which is situated to the left of the ventricle receives blood from a small sinus venosus. There is no conus arteriosus. The ventral aorta is slightly swollen at the base to form *bulbus arteriosus*. The ventral aorta gives off paired *afferent branchial arteries* bring deoxygenated blood and *efferent*

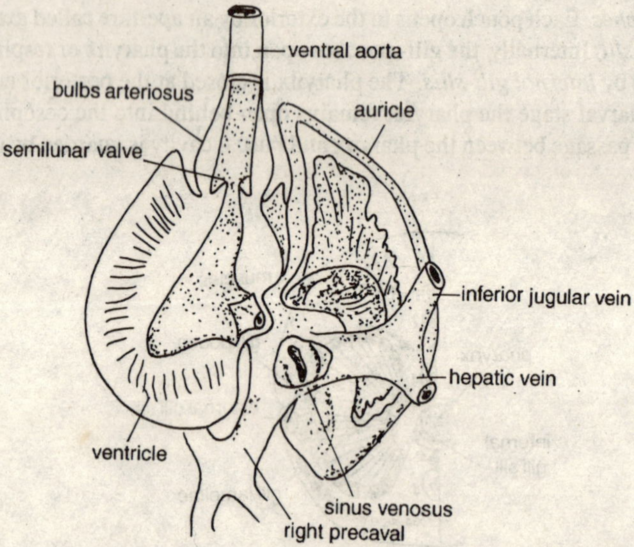

Fig. 2.10. *Petromyzon*. Heart.

branchial arteries collect the oxygenated blood from the gills and convey it to the dorsal aorta. Blood from the body is collected by a pair of *cardinal veins* which are continuation of the *caudal veins*. There is no renal portal system, the blood from the kidneys drains into two *sub-cardinal sinuses* which are connected by numerous small passages with the cardinal veins above. A pair of *jugular veins* and a median ventral *inferior juglar vein* collect blood from the head region.

The *left precaval* is disappered in adult, so that the jugular and cardinals of both the sides open into the right precaval. A *hepatic portal system* is present. The hepatic vein opens in sinus venosus.

The red blood corpuscles or *erythrocytes* of blood contain a respiratory pigment, the *haemoglobin*. The *haematopoiesis* occurs primarily in the spiral valve. In adult, this function is shared with the reticular tissue of the kidney. The erythrocytes are nucleated and oval.

In addition to the blood-vessels, the circulatory system contains lymph vessels or *lymphatics*.

In the *Petromyzon* spacious *lymph-sinuses* surrounding the blood vessels are present, and there are communications between the lymphatics and the coelom by means of minute apertures called *stomata*. The spleen is absent.

NERVOUS SYSTEM

1. Central Nervous System

Brain. The brain of *Petromyzon* is in a relatively primitive condition. The brain conforms to the typical plan of vertebrate brain. The fore-brain consists of *olfactory bulbs*, *cerebral hemispheres*, and *diencephalon*. The olfactory bulbs are large and widely separated. The cerebral hemispheres are very small and smooth structures and usually smaller than the olfactory bulbs. The diencephalon is large. On the dorsal side the diencephalon bears two unequal *Habenular ganglia* ; the right being much larger than the left. These constitute an epithalamus and are connected to the *pineal apparatus*. This consists of the *pineal eye* (photosensitive) and the small *parapineal body* lying below it and connected with the left Habenular ganglion. The pineal eye consists of a pigmented retina and a flat imperfect lens with sensory cells. On the ventral side, a well developed *hypothalamus* is present. It has small *lobi inferiores*, indistinct *saccus vasculosus* and prominent *infundibulum* bearing *hypophysis* or *pituitary gland*.

The midbrain is represented by *optic lobes* and *crura cerebri*. The optic lobes are very imperfectly differentiated in the tectum of the mid brain.

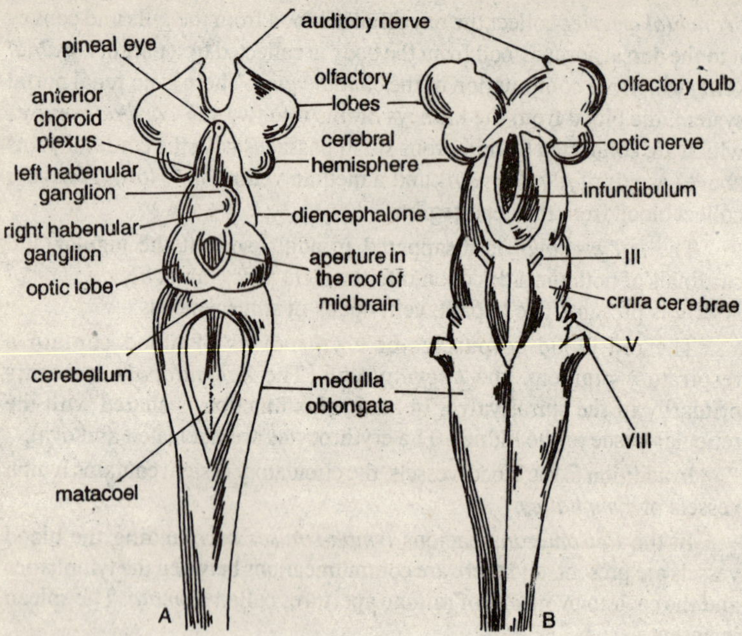

Fig. 2.11. *Petromyzon*. Brain A—Dorsal view, B—Ventral view.

The hind-brain comprises the *cerebellum* and *medulla oblongata.* The cerebellum is very small and occurs as a mere transverse band roofing over the anterior end of the fourth ventricle. The medulla oblongata is highly.developed. It has giant cells which are connected to the muscles of tail, by large prominant *Muller's fibres* to control the movements of tail.

The mid brain has *middle choroid plexus* which is absent in higher vertebrates where such structures are limited to the fore-brain and hind-brain as anterior and posterior choroid plexuses.

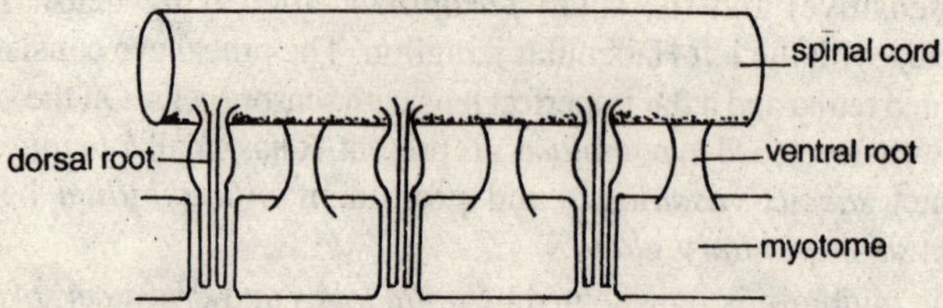

Fig. 2.12. *Petromyzon*. Spinal cord and Spinal nerve.

Spinal Cord. It is dorso-ventrally flattened, transparent and nonvascular tube. It is made up of grey matter only, the white matter present in higher vertebrates, is absent here. It encloses a narrow *central canal*. The dorsal septum and ventral fissure are absent.

2. Peripheral Nervous System

It comprises of *cranial nerves* and *spinal nerves*. From the brain 10 pairs of cranial nerves are given out. The spinal nerves arise from the spinal cord have separate dorsal and ventral roots.

3. Autonomic Nervous System

It is present in the form of *intestinal plexuses* which are connected to the brain. The autonomic nerve fibres are present in all the spinal nerves. The sympathetic nervous system is poorly developed.

SENSE ORGANS

In *Petromyzon* the organs of smell, sight and hearing are present.

Organ of Smell

There is a median *external naris* present on the dorsal side of the brain. It leads by a short passage into a rounded *olfactory sac*. The sac is present just infront of the brain and having its posterior wall raised into ridges covered by olfactory mucous membrane. The membrane have olfactory receptor cells which are supplied with the olfactory nerve fibres. Water is drawn into the sac for the testing and then expelled out through the external naris. From the bottom of olfactory sac is given off a large *naso-hypophysial sac* which extends between the brain and the floor of skull and acts as rubber of a pipette.

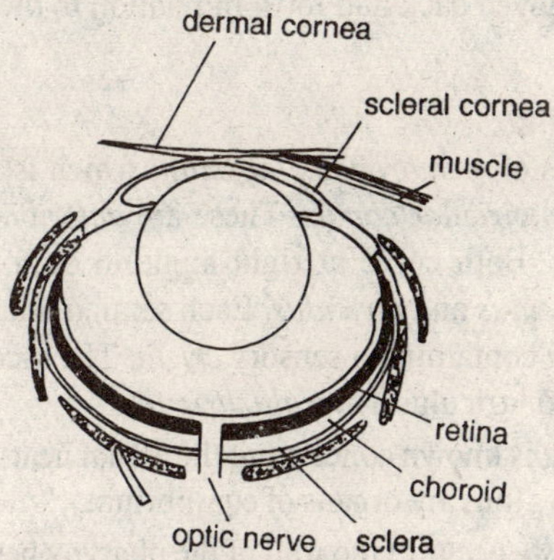

Fig. 2.13. *Petromyzon*. The eye.

Organs of Sight

The eyes are very well developed and are movable with the help of six eye muscles. The eye-lids are absent. The eyes are covered by transparent skin, the *conjunctiva*. The *sclerotic* is membraneous. The *retina* has more rods than the cones. The *lens* is spherical and never changes its shape. The *suspensory ligament* is absent hence the lens is held in position by vitreous humour.

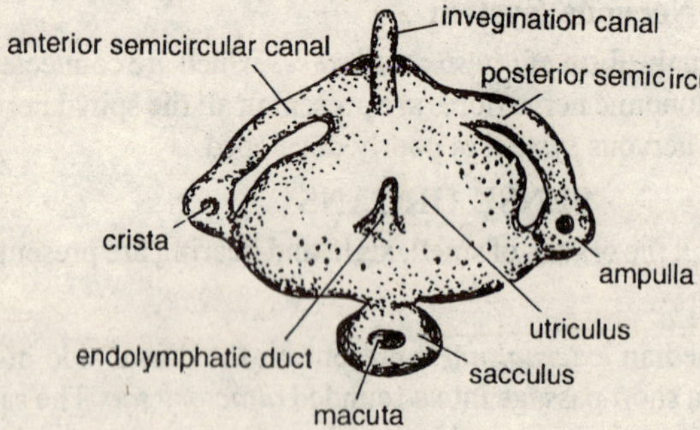

Fig. 2.14. *Petromyzon*. Membranous labyrinth.

The accommodation is brought about by *cornealis muscle*, which arises from the specialization of two anterior myotomes, pulls on the cornea from one side and flattens it. This corneal movement thrusts the lens in towards the retina. The whole lens moves in a manner roughly to the lens of a camera being moved back and forth in relation to the photographic film.

Organs of Hearing

The ear consists of *membranous labyrinth* which is remarkable in having only two *semicircular canals*. These are *anterior* and *posterior semicircular canals.* Both curve at right angle to each other and are equipped with a *sacculus* and *utriculus*. Each semicircular canal has, at one end, an *ampulia* containing a sensory *crista*. The sacculus contains a *macula sacculi* and utriculus a *macula utriculi*.

Little or nothing is known concerning the actual hearing capacity in *Petromyzon*. The ears thus are organs of equilibrium.

The *organ of taste* occur in the wall of the pharynx between the gill-sacs.

Lateral line-System

The lateral line sense organs are simple and are exposed to the surrounding water. They occur in the head region and in a single line extending along each side of the body.

URINOGENITAL SYSTEM

Excretory System

The *excretory system* is for the first time appeared as kidneys in *Petromyzon*. These are long strip-shaped bodies developed from *mesonephros* of the embryo remain attached to the dorsal body wall by peritoneum. Each kidney is made up of numerous minute renal units known as *Malpighian corpuscles*. Each corpuscle contains a bunch of blood vessels, the *glomerulus,* and opens into a winding tubule.

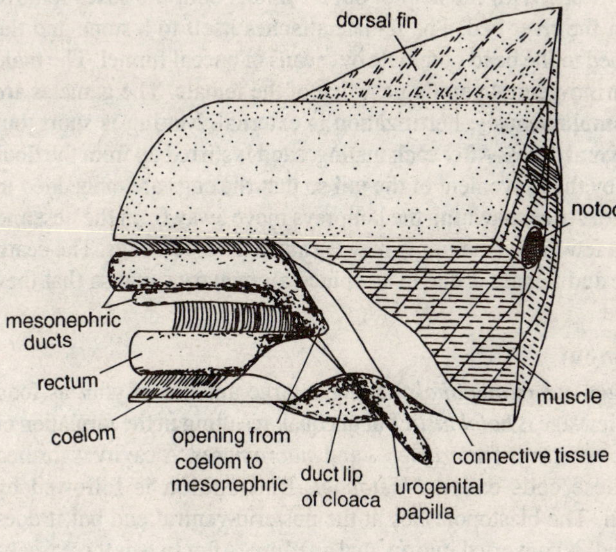

Fig. 2.15. *Petromyzon*. Cloacal region.

Along the free edge of kidneys run the *mesonephric ducts* which open posteriorly into small urino-genital sinuses placed just behind the rectum. The urinogenital sinus has a pair of genital pores in its lateral walls and opens by a conical urinogenital papilla into the *cloaca*. The cloaca is present on the ventral side at the junction of the trunk and tail.

Reproductive System

The sexes are separate in the adult but sexual differentiation occurs at a late stage so that both *oocytes* and *spermatocytes* may be present in the same gonad in young *Petromyzon*. There is a single gonad lies between the kidneys. It remains suspended from dorsal body wall by a fold of peritoneum, the *mesorchium* in the male and *mesovarium* in the female. The genital ducts are absent. The ova as well as the spermatozoa are liberated into the coelom by the rupture of gonad. The gametes escape through a pair of genital pores. Thus the fertilization is external. In adult male *interstitial* or *Leydig-cells* occur between the follicles. The follicles produce sperms.

Breeding

Petromyzon breeds only once in its whole life. It migrates from sea and enters the fresh water for spawn. During Spring or early Summer they collect in pairs in shallow spawning grounds in the river and the female prepares a 'nest'. With the help of buccal funnel both the sexes remove stones from the river bed. The female attaches itself to a stone and the male attached to the head of female by means of buccal funnel. The male brings its urino-genital papilla near that of the female. The gametes are extruded simultaneously. Fertilization is external. Mating is short, but repeated several times. After each mating, sand is stirred up from the floor of the nest by the movement of the tail so that the eggs are embedded in settling sand. After spawning the lampreys move away from the nest and die within a few days. A lamprey may lay about 2,35,000 eggs. The death of the male and female is due to atrophied alimentary canal so that they can not feed.

Development

The eggs are *telolecithal*, having a large amount of yolk as food material. Cleavage is *holoblastic* but unequal, resulting in the formation of a blastula consisting of *micromeres* and *macromeres*. A cavity is formed between these cells called *blastocoel*. Blastutation is followed by gastrulation. The blastopore lies at the posterio-ventral end but it does not close and, is converted into an anal aperture. After two or three weeks of fertilization an *ammocoete larva* comes out of the egg.

Ammocoete larva

The ammocoete is about 8-10 mm long, elongated and worm-like. It differs from the adult in several respects.

There is a semi-circular, hood-shaped *upper-lip* instead of the suctorial buccal funnel present in adult. The *lower lip* is transverse. Teeth are absent but several branched buccal tentacles surround the mouth. A *velum* consisting of two cup shaped flaps separate the buccal cavity from pharynx.

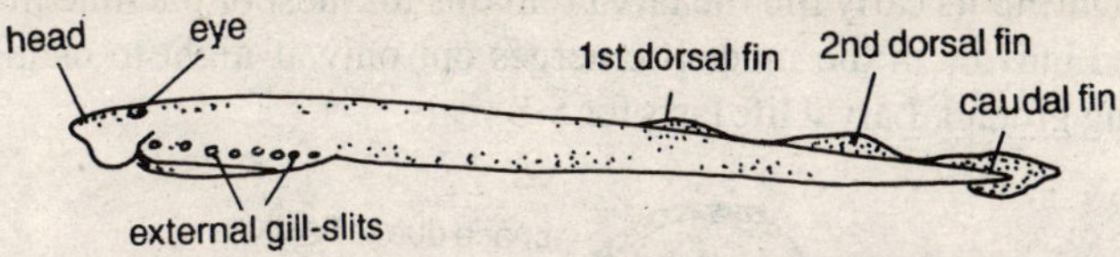

Fig. 2.16. Ammocoete larva.

The eyes are vestigeal and covered over by the skin. The parietal eye is well-developed. The paired fins are absent only a median fin is continuous. Seven pairs of gill-sacs are present on the lateral sides of the trunk. The gill-sacs internally open into the pharynx. The *endostyle* and *peripharynegeal grooves* are present in alimentary canal. The animal is filter-feeder. The endostyle secretes mucuos that passes forward along a groove in the floor of pharynx to a pair of peripharyngeal grooves behind the velum. The microscopic food particles which come along with water current entangled in mucous. This mucous is finally transported to the intestine by the cilia of *hyperpharyngeal groove*. There is a small liver, a bile-duct and a gall-bladder which disintegrate during metamorphosis. Active pronephric kidneys are present. The skull and the branchial basket are rudimentary.

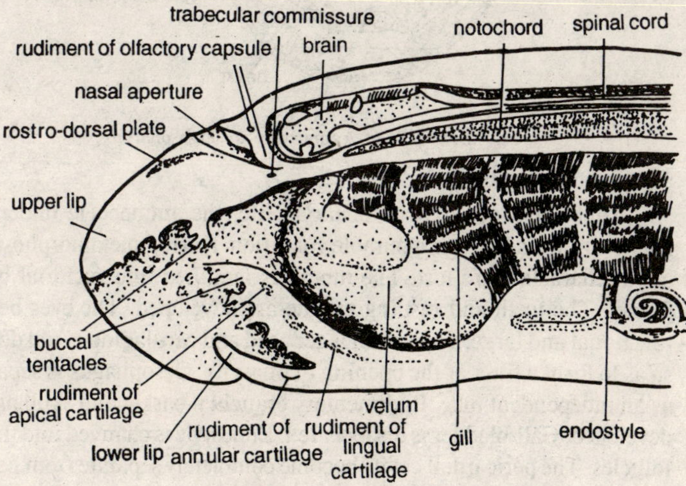

Fig. 2.17. Ammocoete. Section of anterior region.

During its early life the larva remains for most of the time in a 'U' shaped burrow in the mud. It emerges out only at night to change its feeding ground. Larval life lasts for 5-8 years.

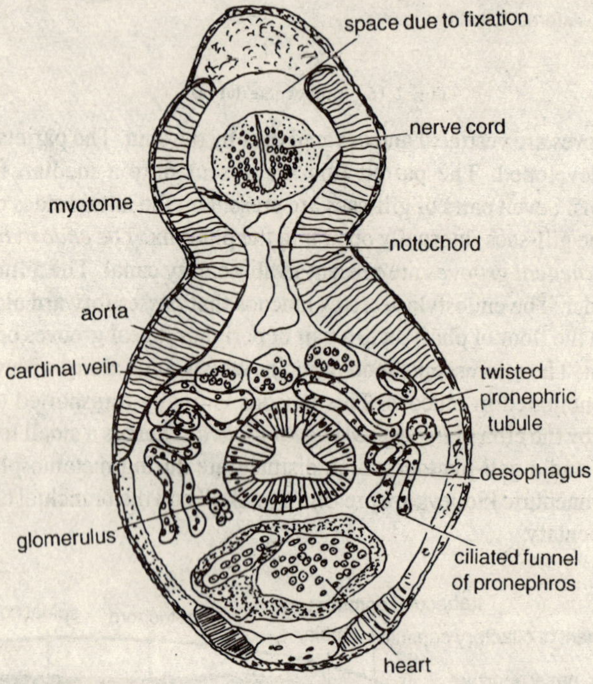

Fig. 2.18. T.S. of Ammocoete through pharynx.

Metamorphosis

After about seven years of larval life, the ammocoete under goes metamorphosis during favourable condition. During metamorphosis few larval characters are lost. The mouth is modified into suctorial buccal funnel. The teeth and rasping tongue are developed. The eyes become functional and large. The velum is reduced and cartilaginous rod develop so as to form a filter at the opening of pharynx. Oesophagus is separated as an independent tube. Rudimentary branchial basket and skull are full developed. Gall-bladder is disappeared. Endostyle is changed into thyroid follicles. The pericardial cavity become completely separate from general body cavity. The pronephros of larva disappears and the mesonephric kidneys are formed. The round spinal cord of larva becomes flattened and neural arches appear in the trunk region of adult. The head muscles are reformed. The skin colour is also changed from muddy brown to metallic. The continuous median fin becomes broken into two dorsal and one caudal fin. Now the adult comes back to the sea from river and leads a sanguivorous life.

Differences between *Petromyzon* and *Myxine*.

Petromyzon	*Myxine*
1. These are marine as well as fresh water	1. These are exclusively marine.
2. They are ectoparasite. They do not bore into the prey but attach to them.	2. They are endoparasite as they bore into their preys.
3. Rasp away flesh and sucks the blood of prey (sanguivorous).	3. Primarily scavengers as they burrow into dead or morbid fishes for their flesh.
4. The shape of body is cylindrical and stout measuring upto one metre.	4. The body shape is eel-like and feeble measuring under one metre.
5. The dorsal fin is divided into two by a notch. The caudal fin is well-represented.	5. The dorsal fin is single or absent The caudal fin is slightly developed.
6. Less slimy.	6. Exceedingly slimy due to presence of numerous mucous glands.
7. Mouth is ventral or subterminal in position due to presence of buccal funnel.	7. Mouth is terminal in position and is surrounded by soft lips. Buccal funnel is absent.
8. Tentacles around mouth are absent	8. Three to four pairs of oral tentacles present around the mouth.
9. Tongue is feebly developed with larger horny teeth.	9. Tongue is strongly developed with smaller teeth.
10. Single dorsal-median nostril present high on the head.	10. Single median nostril lying very close to mouth and opens terminally.
11. Paired eyes are well-developed and functional.	11. Paired eyes are rudimentary and concealed under skin. The lens and eye-muscles are absent.

12. The pineal eye is present.	12. The pineal eye is absent.
13. Paired salivary glands are present which secrete anticoagulant to prevent coagulation of host's blood.	13. Paired salivary glands are absent.
14. Pharynx ends blindly as a respiratory tube.	14. Pharynx opens into oesophagus.
15. Gill-sacs are seven pairs that open into the respiratory tube, not in gut.	15. The number of gill-sacs varies, six pairs in *Myxine* and 13 - 1 5 pairs in *Eptatretus*. The gill-sacs open into the gut.
16. Seven pairs of external gill-slits.	16. One pair in *Myxine* and 13-15 pairs in *Eptatretus*.
17. A spiral valve or typhlosole is present in intestine.	17. Intestine possesses longitudinal folds.
18. The skull is imperfectly roofed.	18. The skull is without roof.
19. The neural arches are present round notochord.	19. The neural arches are absent.
20. Well developed branchial basket around the gill-sacs.	20. Small and poorly developed branchial basket.
21. A poorly developed lingual cartilage to support the tongue.	21. Well developed lingual cartilage is present.
22. The pericardial sac is thick walled and is supported by pericardial cartilage.	22. The pericardial sac is thin walled and is not supported by cartilage.
23. Each aortic arch divides and send branches to the posterior hemibranch and anterior hemibranch of adjacent gill-sacs.	23. Each aortic arch supplies the hemibranch of a single gill-sac.
24. The mesonephros is functional adult kidney.	24. The pronephros is retained in the adult. The functional adult kidney is mesonephros which is simple.
25. The gall-bladder and bile-duct are absent in adult.	25. Both are present in adult.
26. The nasohypophysial duct ends blindly.	26. It opens behind into pharynx.

27. The brain is more developed. Its fore-brain and mid-brain are distinct.	27. The brain is rudimentary its fore-brain and mid-brain are not distinct.
28. Ten pairs of cranial nerves.	28. Eight pairs of cranial nerves.
29. The spinal nerves have separate dorsal and ventral roots.	29. The spinal nerves have united dorsal and ventral roots.
30. Two semi-cirular canals are present in ears.	30. True ears are absent but one semicircular canal is present.
31. The lateral-line system is much better developed.	31. The lateral-line system is poorly developed.
32. The urino-genital sinus is present.	32. It is absent.
33. The sexes are separate but differentiated in the later stage of adult.	33. Sexes are united. The anterior part of gonad acts as ovary and the posterior as testis.
34. For breeding and spawning they ascend fresh water rivers and streams i.e. *anadromous.*	34. Spawning occurs on ocean bed.
35. The eggs are small, without shell and are telolecithal.	35. The eggs are large and enveloped by horny shell. They are telolecithal.
36. Segmentation is holoblastic but unequal.	36. Cleavage is meroblastic in *Eptatretus.*
37. Development indirect with an ammocoete larva that undergoes metamorphosis.	37. Development is direct without larva and the metamorphosis.

AFFINITIES OF CYCLOSTOMATA

Cyclostomes represent the most primitive members among living vertebrates. Ammocoete larva illustrates the primitive characters like Cephalochordata while the adult has certain specialized as well as degenerate features due to parasitic mode of life. Thus the affinities of cyclostomes can be discussed under following headings :

1. Primitive Characters

(a) Resemblances with Cephalochordata

1. Head is indistinct.
2. Skin naked without any exoskeleton. Paired fins are absent.

3. Numerous gill-slits are present.
4. Persistent. and continuous notochord.
5. Segmental muscles or myotomes with little modification from anterior to posterior end.
6. Alimentary canal is straight.
7. Separate dorsal and ventral roots of spinal nerves. Nerves are unmyelinated.
8. Gonoducts are absent. Fertilization external.
9. Development indirect with a larval stage. The ammocoete larva resembles *Amphioxus* in; continuous median fin; presence of endostyle; filter feeding; ciliated gut; and vestibule anterior to mouth.

(b) Differences from Fishes

1. Absence of jaws, scales, true teeth, paired appendages, true fin-rays, girdles, ribs, stomach, spleen, gall-bladder and gonoduct.
2. Caudal fin *diphycercal*.
3. Continuous median dorsal fin.
4. Incomplete cranium and vertebrae.
5. A single median nostril.
6. Poorly developed pancreas, brain, spiral valve, sympathetic nervous system and lateral line system.
7. Heart 'S' shaped without conus arteriosus.
8. 9th and 10th cranial nerves are not enclosed by cranium. Only 8 pair are present in *Myxine*.
9. One or two semi-circular canals in membranous labyrinth.

(c) Affinities with Ostracoderms

Cyclostomes resemble the ostracoderms in the absence of jaws, structure of brain, pouch like gill-sacs, arrangement of cranial nerves, single nasal opening, internal ear with two semicircular canals and lateral line system.

2. Vertebrate Characters

1. Head with paired sense organs-like eyes and ears.
2. Differentiated brain with 10 pairs of cranial nerves.
3. Cranium for the protection of brain.
4. Beginning of segmental vertebrae.
5. Dorsal root ganglia on spinal nerves.
6. A large bilobed liver without gall-bladder and bile duct.
7. Gills for respiration.

8. Blood contains erythrocytes and leucocytes.
9. Well-developed circulatory system with a heart and lymphetic system is also present.
10. Mesonephric kidneys.
11. Lateral-line sense organs.
12. 'E' shaped myotomes as in fishes.

3. Specialized Characters

1. Suctorial mouth is present at the bottom of funnel which helps in leading a parasitic life to the animal.
2. Powerful, muscular tongue bearing teeth.
3. Peculiar sac-like gill-chambers located far behind the head.
4. Complete separation of ventral sac-like respiratory pharynx from the dorsal oesophagus.
5. A special supporting skeletal structure called the branchial basket is present around the gills in the pharyngeal region.
6. A definite liver is present.
7. A single median naris is present.
8. Large number of mucous glands are present in *Myxine*.

4. Degenerative Characters

1. Body is elongated and eel-like.
2. Paired fins and the girdle which support them are absent.
3. The exoskeletal structures such as scales are absent.
4. Liver is present but gall-bladder and bile duct are absent.
5. The eyes are poorly developed and without eye-lids i.e. concealed under skin.

Thus looking at the characters, described above, shows that cyclostomes have developed a number of characters which are due to parasitic mode of life. They differ widely from the fishes in certain characters, like in the absence of biting jaws, absence of paired fins, presence of rasping tongue etc. Some authors considered cyclostomes as degenerate fishes due to absence of jaws, limbs and girdles. Hence, cyclostomes were separated from the fishes and placed in a separate group Agnatha. Agnatha includes the extinct ostracoderms and living cyclostomes. According to *Stensio* (1927) the cyclostomes have descended from the ostracoderms by the evolution of a sucking mouth, replacement of bony skeleton by cartilage and loss of paired limbs.

There is increasing evidence to show that there were two major groups of ostracoderms; the *cephalaspids* having several pairs of gill-slits, paired fin-like appendages, heterocercal tail and a median single nostril and the *pteraspids* having a single pair of gill-slits, unpaired appendages, hypocercal tail and two separate nostrils. According to *Stensio* the lampreys represent derivatives of cephalaspid line and the hagfishes are probably derived from pteraspid line. Thus this view holds that cyclostomes are diphylatic in origin.

A new idea put forwarded by *Jaravick*, according to him the cyclostomes and gnathostomes are sister group and that the gnathostomes betray several features even more primitive than the earliest recorded cyclostomes. Thus the systemic position of cyclostomes remains doubtful.

Revision Question

1. Describe the geographical distribution, habits and external features of *Petromyzon*.
2. Give a detailed account of external features of *Petromyzon*.
3. Describe the struture of *Petromyzon*. Discuss the differences between *Petromyzon* and *Myxine*.
4. Write short notes on : (a) Ammocoete larva, (b) Branchial basket, (c) Buccal funnel.

3

Pisces : Characters and Classification

Characters

1. These are the vertebrates adapted for a life in water.
2. Body is spindle-shaped, stream-lined and the tail is propulsive organ in water.
3. Paired appendages in the form of extension of skin are known as fins. Fins are supported by fin-rays.
4. Respiration, chiefly by gills.
5. Exoskeleton of dermal scales.
6. Lateral-line sensory system is present.
7. Ears are mainly for equilibrium.
8. Sexes are separate.

CLASSIFICATION

Pisces is divided into 4 classes : Placodermi, Chondrichthyes, Osteichthyes and Dipnoi.

Class-I. Placodermi

1. These are extinct fishes. They flourished from early Devonian to Permian periods.
2. Body was typically fish-like.
3. They had a bony armour like the ostracoderms.
4. Jaws were not properly formed.
5. Paired and unpaired fins were present.
6. Cranium was partly ossified.
7. Vertebral column had neural and haemal arches.
8. Notochord persisted throughout life.

Class - Placodermi is divided into 6 sub-classes or orders:

Sub-class 1 : Acanthodii

Example: *Climatius.*

Sub-class 2 : Arthrodira

Example: *Coccosteus.*

Sub-class 3 : Petalichthyida

Example : *Macropetalichthys.*

Sub-class 4 : Antiarchi

Example : *Petrichthys.*

Sub-class 5 : Rhenanida

Example : *Gemuendina*

Sub-class 6 : Paleospondyli

Example: *Paleospondylus*

Class-II. Chondrichthyes or Elasmobranchii

1. Body is laterally compressed and spindle shaped.
2. Paired and unpaired fins are present.
3. Endoskeleton entirely composed of cartilage, sometimes calcified.
4. Exoskeleton consists of dermal denticles or *placoid scales*.
5. Jaw suspension is *holostylic*, *amphistylic* or *hyostylic*.
6. Gills are never covered by opercula except some.
7. Tail is always *heterocercal*.
8. No lungs or air bladders exist.
9. Males usually have *claspers*.
10. They may be oviparous, *ovo-viviparous* or *viviparous*.
11. Kidneys are *mesonephric*.

Class -Chondrichthyes is divided into two sub-classes : Selachii and Holocephali.

Sub-class-1. Selachii

1. Skin is covered by placoid scales.
2. Gill-slits are uncovered. They are 5-7- pairs.
3. Jaw suspension hyostylic or amphistylic, and numerous teeth arranged in several rows.
4. Claspers occur in association with the pelvic fins.

Sub-class. Selachii is divided into two orders :

Order (i) Pleurotremata or Squaliformes

1. Gill-slits open on the lateral sides of the body.
2. Spiracles, when present, lie behind the eyes.
3. Pectoral fins are distinctly marked off from the body.

4. They are surface feeder.

   Example : *Scoliodon, Squalus, Sphyrna, Rhineodon* etc.

Order (ii) Hypotremata or Rajiformes

1. Body is dorso-ventrally flattened.
2. Gill-slits are present on ventral side.
3. Spiracles are usually present.
4. Pectoral fins are not well-marked.
5. They are bottom feeders.

   Examples : *Pristis, Torpedo, Raja, Trygon* etc.

Sub-class-2. Holocephali

1. Body is laterally compressed.
2. Mouth is small bounded by lips.
3. Skin without scales in adult.
4. Large toothed plate attached to the jaws.
5. Jaw-suspension is holostylic.
6. Gills covered with operculum.
7. Extra clasper in male.
8. Poorly developed vertebrae.

   This sub-class includes only one order Chimaeriformes.

   Example : *Chimaera* (*Hydrolagus*).

Class III. Osteichthyes or Teleostomi

1. Body is usually spindle shaped.
2. Median and paired fins are present.
3. Skin with mucous glands. Exoskeleton of *genoid*, *cycloid* or *ctenoid* scales.
4. Endoskeleton is partly or wholly bony.
5. Notochord often persisted in greatly reduced form.
6. Mouth is terminal or sub-terminal. Nostrils are located on the dorsal side of snout.
7. Gill-slits are covered with *operculum* and the inter-branchial septa are reduced.
8. Jaw suspension is autostylic type.
9. Tail is homocercal, gephycercal or diphycercal.

   Class-Osteichthyes has been divided into two sub-classes : (1) Choanichthyes, and (2) Actinopterygii.

Sub-Class-1. Choanichthyes or Crossopterygii

1. Single or double air-bladder.
2. Fins are paired and lobed.
3. Internal nares present.
4. Skeleton of the pectoral fins with a jointed central axis bearing radials on two sides.
5. Almost all are fresh water fishes. Occur from Devonian to Recent period.

The sub-class-Crossopterygii is divided into two super-orders (i) Osteolepides and (ii) Coelacanthi.

Super Order (i) Osteolepides

1. All are fossil forms.
2. Tail was heterocercal.
3. Body was fusiform and covered with thick bony scales.
4. Scales were cycloid.

It comprises three orders : Osteolepiformes, Holoptychiformes and Rhizodontiformes.

Examples : *Osteolepis, Rhizodus, Strepsodus* etc.

Super order (ii) Coelacanthi

1. The tail-fin diphycercal consists of three lobes.
2. Only external nares present.
3. Air bladder present.
4. Scales are cycloid.
5. No vertebral centra are present except in the caudal region of some foms.

This super order includes only one order the Coelacanthiformes, with only one living speciman.

Examples : *Latimeria* (Living fossil), *Coelacanthus* (Extinct) etc.

Sub-Class 2 Actinopterygii

1. Commonly known as ray finned fishes. The paired fins have broad base and fan-like form.
2. They lack fleshy lobe and are supported by radiating dermal rays.
3. Olfactory sacs do not open into buccal cavity, but double external nares are present.

Sub class Actinopterygii is usually divided into three infra classes Chondrostei, Holostei and Teleostei, But there are certain controvercy regarding their division. So it is better, not to divide Actinoplerygii into these infra-classes but to classify into a series of orders. The whole sub-class is divided into two groups which contain 60 orders.

Group A This group has been separated to accomodate the Polypteriformes which differ from all the Actinopterygii in the structure of pectoral endoskeleton and of the dorsal fin.

Order-Polypteriformes

1. The dorsal fin is split into numerous finlets.
2. Body is elongated and covered over by rhombic or ganoid scales that do not overlap.
3. Tail is very short and carries a rounded nearly symmertical fin.
4. A double air bladder connected with the pharynx by a duct is present, but it is not used as a lung.

   Example : *Polypterus.*

Group B. The Actinopterygii in which the pectoral radials attached at least part, immediately to the scapulo-coracoid cartilage or the scapula and coracoid. According to *Berg*, Group B is divided into 59 orders.

1. Tarrasiiformes
2. Palaeonisciformes
3. Gymnonisciformes
4. Phanerorhynchiformes
5. Dorypteryformes
6. Bobasatraniiformes
7. Redfieldiiformes
8. Perlebidiformes
9. Ospiiformes
10. Pholidopleuriformes
11. Saurichthyiformes
12. Acipenseriformes
13. Amiiformes
14. Aspidorhynchiformes
15. Pyenodontiformes
16. Pachycormiformes
17. Lepidosteiformes
18. Pholidophoriformes
19. Clupeiformes
20. Bathyclupeiformes

31. Beloniformes
32. Gadiformes
33. Macruriformes
34. Gasterosteiformes
35. Syngnathiformes
36. Lampridiformes
37. Cyprinodontiformes
38. Phallostethiformes
39. Percopsiformes
40. Stephanoberyciformes
41. Beryciformes
42. Zeiformes
43. Mugiliformes
44. Polynemiformes
45. Ophiocephaliformes
46. Symbranchiformes
47. Perciformes
48. Dactylopteriformes
49. Thunniformes
50. Pleuronectiformes

21. Galaxiiformes
22. Scopeliformes
23. Ateleopiformes
24. Giganturiformes
25. Saccopharygiformes
26. Mormyriformes
27. Cypriniformes
28. Anguilliformes
29. Halosauriformes
30. Notacanthiformes
51. Kosteiformes
52. Chaudhriformes
53. Mastacembeliformes
54. Echeneiformes
55. Tetradontiformes
56. Gobiesociformes
57. Batrachoidiformes
58. Lophiiformes
59. Pegasiformes

Here a descriprtion of only the important ones is given :

Order- Acipenseriformes

1. The body is shark-like and covered with five rows of bony scutes or naked.
2. The head is produced into a flat triangular rostrum bearing barbels.
3. Caudal fin is heterocercal.
4. Spiracles are present between eyes and nares

Example : *Acipenser* and *Polyodon* (paddle fish) etc.

Order-Amiiformes

1. Head bears large bony plates
2. One pair of small barbels above the mouth.
3. Opercula are very large.
4. Caudal fin abbreviate hererocercal

Example *Amia* (Bowfin), *Sinamia* (Extinct).

Order- Lepidosteiformes

1. Body spindle like covered with thick, rhomboid, close fitting scales.
2. Jaws are prolonged into a tapering beak armed with sharp teeth.
3. Caudal fin is rounded and nearly symmetrical (hemiheterocercal).
4. Air bladder cellular.

Example : *Lepidosteus* (The gar-pike)

Order-Clupeiformes

1. No Weberian ossicles.
2. Scales usually cycloid.
3. Caudal fin homocercal.
4. Upper jaw usually bordered by premaxillae.

5. There is a single dorsal fin and a small ventral fin which may even be absent.

   Examples : *Hilsa hilsa*, *Salmo* (Trout), *Notopterus*, *Chirocentrus* etc.

Order-Cypriniformes

1. A peculiar Weberian apparatus connecting the ear with the air bladder is present.
2. Air bladder connected with the gut by means of a duct.
3. Fins either without spines or dorsal, anal and pectoral fins have a spine each.
4. Ventral fins are abdominal some possess an adipose fin.

   Examples : *Lebeo*, *Barbus*, *Cirrhina* etc.

Order-Scopeliformes

1. Photoreceptors are present.
2. No spines in dorsal and anal fins.
3. An adipose fin is present.

   Example : *Harpodon nehereus* (Bombay Duck).

Order-Anguilliformes

1. The body is snake-like without pelvic fins and with reduced or no scales.
2. Dorsal and anal fins long narrow. No spines in the fins.
3. Air bladder connected with the intestine by a duct.

   Examples : *Anguilla*, *Muraena*, *Ophichthys*, etc.

Order-Beloniformes

1. Body elongated covered with cycloid scales.
2. Fins without spines.
3. Ventral fins abdominal.
4. Some of them having capability to glide with the help of enlarged pectoral fins.

   Examples : *Exocoetus*, *Belone* etc.

Order-Gadiformes

1. Ventral fin jugular and no spines in fins.
2. Scales cycloid.

   Example : *Bregmaceros* (Cod fishes).

Order-Syngnathiformes

1. The trunk is invested in bony armour.

2. Mouth is small and lies at the end of a tubular snout.
3. Ventral fin, if present, abdominal or sub-abdominal.

   Examples : *Hippocampus*, *Syngnathus* etc.

Order-Cyprinodontiformes

1. Small sized fishes with ventral fins abdominal.
2. Fins without spines and a single dorsal fin.
3. Lateral line system absent.
4. Lavivorous in habit.

   Example : *Gambusia.*

Order-Mugiliformes

1. Head covered with scales.
2. Scales cycloid or ctenoid. Pelvic fins abdominal.

   Examples : *Sphyraena*, *Mugil* etc.

Order-Polynemiformes

1. Pelvic fins thoracic. Two dorsal fins.
2. Head covered with scales.

   Examples : *Polynemus* and *Eleuthronema.*

Order-Perciformes

1. Two dorsal fins usually with spines.
2. Weberian apparatus absent.
3. Maxillae usually quite excluded from the gap.

   Examples: *Lates*, *Ambassis.*

Order-Pleuronectiformes

1. Flat body adapted for bottom living.
2. Both the eyes are on the upper side.
3. Skull asymmetrical.
4. Air bladder absent in adult.
5. Fins usually without spines.

   Examples: *Pseudorhombus*, *Solea*, *Cynoglossus.*

Order-Mastacembeliformes

1. Fresh water eel-like fishes.
2. Dorsal, caudal and anal fins confluent.
3. Free spines are present before dorsal fin.
4. Buccal cavity enlarged for air breathing.

   Examples: *Mestacembelus*, *Macrognathus.*

Order-Echeneiformes

1. First dorsal fins is modified into an adhesive disc for attachment.
2. Second dorsal and anal fins without spines.
3. Scales cycloid.
4. Air-bladder absent.

   Examples : *Remora* (sucker fish), *Echeneis.*

Order-Tetradontiformes

1. The gill-openings restricted.
2. Ventral fin, if present, thoracic or sub-thoracic.
3. Air sacs present or absent.
4. Air-bladder present or absent.

   Examples : *Tetrodon*, *Diodon*, *Mola* etc.

Order-Lophiiformes

1. Two dorsal fins, the first one, when present transformed into tentacle.
2. Scales obsolete and skin with flaps.
3. Deep sea fishes.

   Example : *Chirolophis.*

Order-Dactylopteriformes

1. Pectoral fins are large.
2. Enlarged keel-like scales on lower part of body.

   Example : *Dactylopterus.*

Order-Batrachoidiformes

1. Body elongated and naked.
2. Head broad and depressed.

   Example : *Batrachus.*

Class-IV. Diponi

1. Dermal bones are present on skull, jaws and pectoral arch.
2. Single gill-slit on either side.
3. Body is covered over by cycloid scales.
4. Internal nares are present.
5. Operculum, cloaca and air bladder present.
6. Paired fins are lobate.

   It is divided into one super-order Ceratodi with two orders:

Order-1. Ceratodiformes

1. Endocranium cartilaginous.

2. Single air-bladder.

   Examples: *Neoceratodus, Ceratodus.*

Order-2. Lepidosireniformes

1. Endocranium membranous.
2. Air-bladder paired.

   Examples: *Protopterus* and *Lepidosiren.*

ZYGAENA OR SPHYRNA

Phylum	-	Chordata
Group	-	Vertebrata
Sub-phylum	-	Gnathostomata
Series	-	Pisces
Class	-	Chondrichthyes
Sub-class	-	Selachii
Order	-	Pleurotremata
Genus	-	*Zygaena*

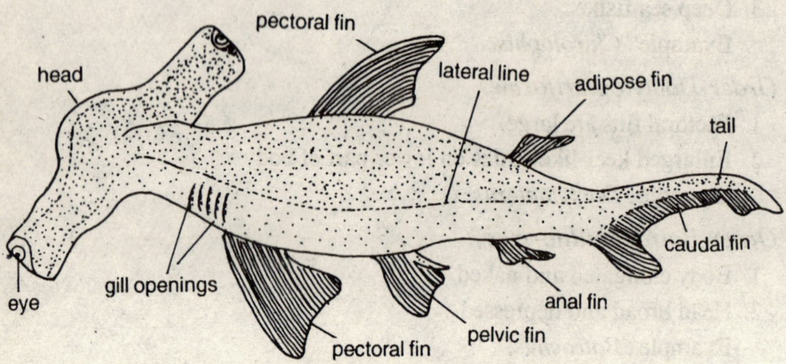

Fig. 3.1. *Zygaena.*

1. It is commonly known as '*hammer headed shark*'.
2. It is a marine fish. It is voracious feeder and eats fishes etc.
3. Body is elongated measuring 4-5 meters and divisible into head, trunk and tail.
4. The head is produced into two prominent lateral lobes supported by corresponding cartilaginous out-growths from postorbital or lateral ethmoidal regions of skull.
5. Eyes situated at the extremites of lobes.

6. Mouth is crescentic and ventral in position.
7. Gill-slits five pairs and lateral in position. No spiracle.
8. Vertebrae asterospondylus.
9. Claspers are found in males only.
10. Viviparous and produces 40 young ones. It is found in all tropical and subtropical sea waters and Pacific Ocean.

TORPEDO

Phylum	-	Chordata
Group	-	Vertebrata
Sub-phylum	-	Gnathostomata
Series	-	Pisces
Class	-	Chondrichthyes
Sub-class	-	Selachii
Order	-	Hypotremata
Genus	-	*Torpedo*

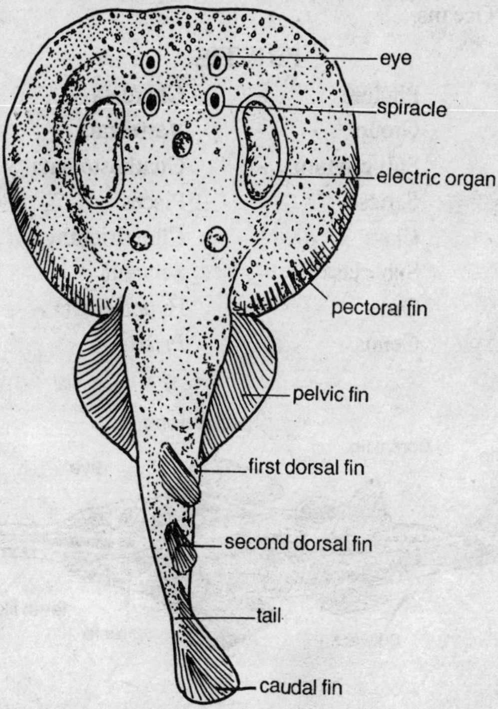

Fig. 3.2. *Torpedo*.

1. It is commonly known as '*electric ray*' because a pair of electric organs are present on either side of the body between head and pectoral fins.
2. It is a marine fish, found on flat, sandy or muddy bottom.
3. Body is regionated into anterior semicircular disk supported by endoskeleton and posterior tail.
4. Two dorsal fins, a caudal fin and two lateral longitudinal folds of pectoral fins present.
5. Large spiracles are situated behind the eyes for respiration.
6. The electric organs are modified lateral muscle plates and are innervated by cranial nerves (VII, IX and X). They produce electric current, the voltage ranges from 8-10 upto 200 volts.
7. The skin is naked. i.e without any scales.
8. Gill-slits present on ventral side.
9. They are viviparous produce young ones.
10. It is present in the Mediterranean Sea, Red Sea and Atlantic and Pacific Oceans.

PRISTIS

Phylum	-	Chordata
Group	-	Vertebrata
Sub-phylum	-	Gnathostomata
Series	-	Pisces
Class	-	Chondrichthyes
Sub-class	-	Selachii
Order	-	Hypotremata
Genus	-	*Pristis*

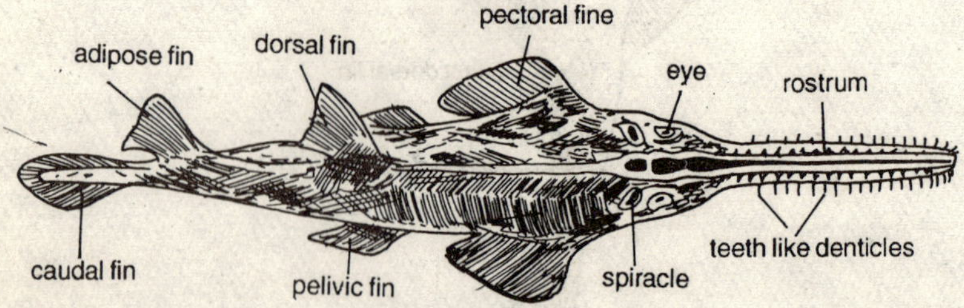

Fig. 3.3. *Pristis.*

1. It is commonly known as '*saw fish*' measuring 3-6 meters in length.
2. It is a warm water, predaceous fish feeding on other fishes and animals. It often ascends the river.
3. Elongated shark-like body differentiated into head, trunk and tail.
4. Head contains one pair of eyes and one pair of spiracles behind eyes.
5. Snout is anteriorly produced into a saw-like rostrum with large and small weakly embedded teeth. Lateral margins contains 16-32 pairs of denticles.
6. The cadual fin is heterocercal.
7. First dorsal fin is opposite to pelvic fin.
8. These are *viviparous.*
9. It is present in all the tropical and sub-tropical seas. It is reported from the America, Gulf of Mexico, Lower Mississippi, Mediterranean Sea and Atlantic Sea.

TRYGON

Phylum	-	Chordata
Group	-	Vertebrata
Sub-phylum	-	Gnathostomata
Series	-	Pisces
Class	-	Chondrichthyes
Sub-class	-	Selachii
Order	-	Hypotremata
Genus	-	*Trygon*

1. It is commonly known as '*sting-ray*' and found lying on the bottom of the sea.
2. It is carnivorous feeds on the small fishes, molluscs and crustaceans.
3. The body is kite-shaped measuring 2 meters long.
4. The skin is naked without scales.
5. Head bears slightly arched mouth on ventral side. Infront of mouth large nares, each divided into two, are present.
6. Five pair of gill-slits on ventral side.
7. Eyes are dorsal and large spiracles are present behind them.
8. Pectoral fins are extensive and meet together infront of the snout.
9. Pelvic fins are smaller and male bears claspers.
10. The tail is very long and bears a serrated sting at the base of the tail.
11. *Viviparous.*

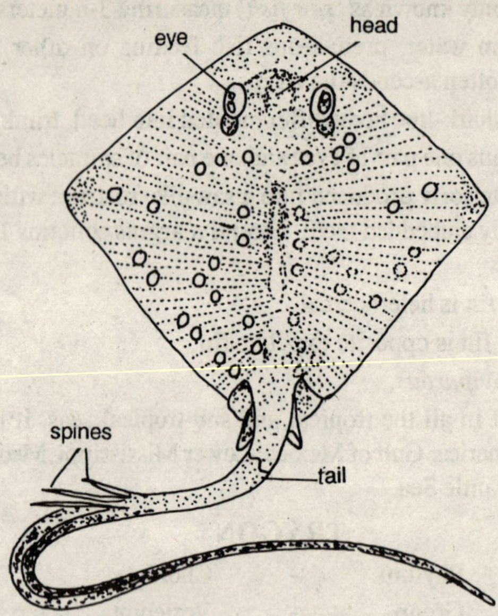

Fig. 3.4. *Trygon.*

12. *Trygon* is found in the Mediterranean, Atlantic, Pacific and Indian Oceans.

CHIMAERA

Phylum	-	Chordata
Group	-	Vertebrata
Sub-phylum	-	Gnathostomata
Series	-	Pisces
Class	-	Chondrichthyes
Sub-class	-	Holocephali
Genus	-	*Chimaera*

1. They are commonly known as '*King of Herrings*', as they are in the habit of pursuing shoals of Herrings. They are also called '*rat-fishes*'.
2. They are marine and measure upto1.5 meter in length.
3. The body is shark-like, but bizzare looking body.
4. The skin is naked without any scales.
5. Head is laterally compressed and has a blunt snout.
6. Mouth is ventral. The nostrils are large and also ventral in positon.
7. There are two dorsal fins and one small ventral fin.

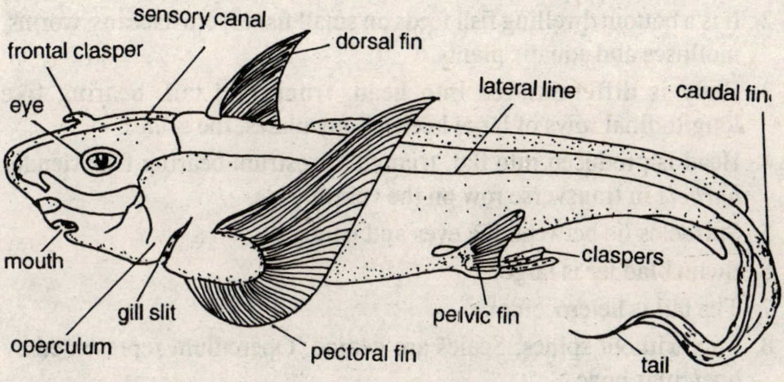

Fig. 3.5. *Chimaera.*

8. Caudal fin is prolonged into a long, rat-like tapering filament.
9. Sexes are separate. Male possess five claspers. A frontal clasper on head, a pair of anterior clasper infront of pelvic fins and a pair of ordinary claspers near the pelvic fins.
10. *Oviparous,* lays heavily yolky eggs with horny cases.
11. It occurs in the Atlantic and Mediterranean Seas.

ACIPENSER

Phylum	-	Chordata
Group	-	Vertebrata
Sub-phylum	-	Gnathostomata
Series	-	Pisces
Class	-	Osteichthyes
Sub-class	-	Actinopterygii
Order	-	Acipenseriformes
Genus	-	*Acipenser*

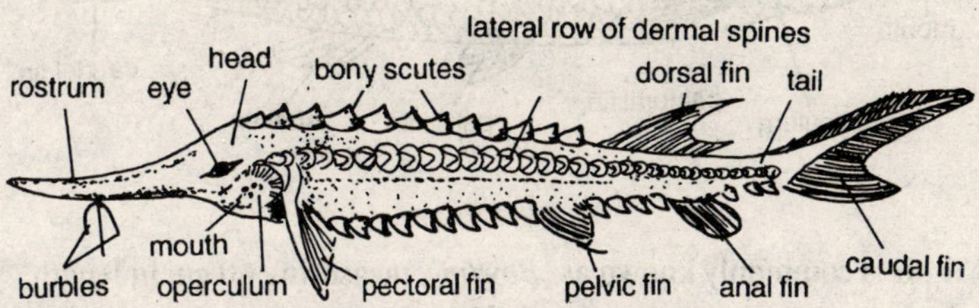

Fig. 3.6. *Acipenser.*

1. It is commonly known as '*sturgeon*', measuring 2-4 meters in length.
2. It is a bottom dwelling fish feeds on small fishes, crustaceans, worms, molluscs and aquatic plants.
3. Body is differentiated into head, trunk and tail, bearing five longitudinal rows of large keeled bony plates, the *scutes.*
4. Head is produced into flat, triangular rostrum bearing four slender *barbels* in transverse row on the ventral side.
5. Spiracles lie between the eyes and nares.
6. Swim bladder is large.
7. The tail is heterocercal.
8. Fins without spines. Scales are geniod. Operculum represented by opercular bone.
9. *Oviparous* 2 to 3 million eggs are laid by single fish during the breeding season.
10. It inhibits rivers and seas of Europe and Asia.

AMIA

Phylum	-	Chordata
Group	-	Vertebrata
Sub-phylum	-	Gnathostomata
Series	-	Pisces
Class	-	Osteichthyes
Sub-class	-	Actinopterygii
Order	-	Amiiformes
Genus	-	*Amia*

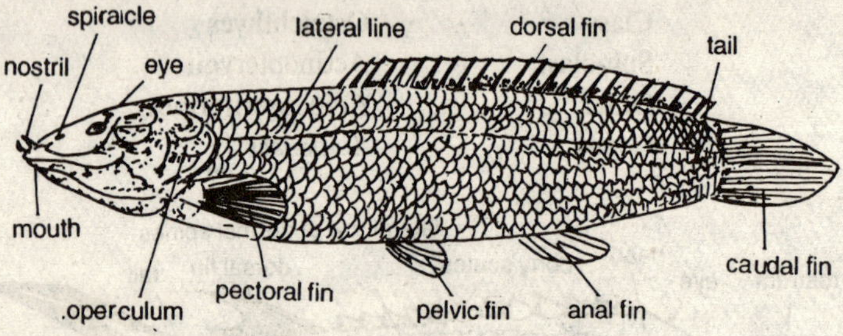

Fig. 3.7. *Amia.*

1. It is commonly known as '*Bowfin*', measuring 60 cm in length.
2. It is present in rivers and lakes feeding on fishes, insects and crustaceans.

3. Body is covered by uniform, thin, overlapping cycloid scales.
4. Mouth is ventral. Nostrils are dorsal. There is a pair of small barbels above the mouth near the nares.
5. Opercula are very large.
6. Dorsal fin has long base and continuous for the greater part of the trunk and tail and contains about 48 fin-rays. All fins are short and devoid of fulcra.
7. Tail is homocercal.
8. The air-bladder is used as lung.
9. Certain primitive characters are present like vestigial spiral valve in intestine; conus arteriosus in heart etc.
10. *Oviparous*. Eggs are laid in nest which are cared by male. Male possess a black spot at the base of caudal fin.
11. It occurs in the rivers of North America.

EXOCOETUS

Phylum	-	Chordata
Group	-	Vertebrata
Subphylum	-	Gnathostomata
Series	-	Pisces
Class	-	Osteichthyes
Sub-class	-	Actinopterygii
Order	-	Beloniformes
Genus	-	*Exocoetus*

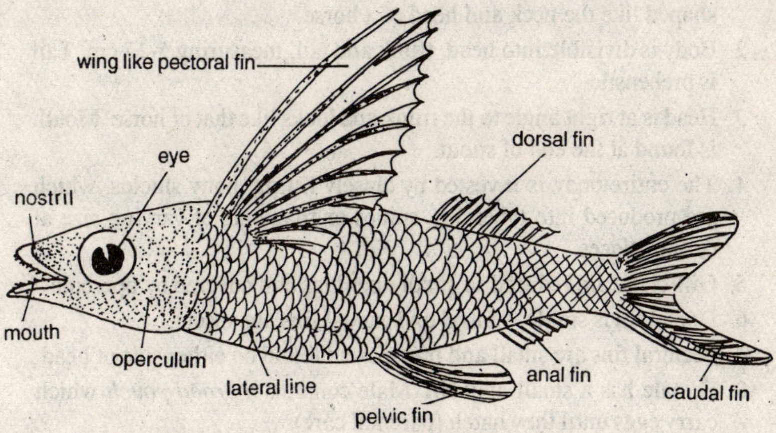

Fig. 3.8. *Exocoetus*.

1. It is commonly known as '*flying fish*', measuring about 30 cm in length.
2. Body is long sub-cylindrical covered with large cycloid scales.
3. The pelvic fins are exceptionally large, spread like wings and make gliding flights. They act as parachutes.
4. The dorsal and anal fins are short and supported by 8-16 fin-rays each.
5. Pelvic fins lie on the ventral side nearly at the middle of the body.
6. Tail is hypoblastic i.e. ventral lobe of tail fin is large.
7. *Oviparous.*
8. Lives in the shoals or groups.
9. It is pelagic and feeding on prawn, fishes and their eggs.
10. It is found in tropical and warmer Atlantic, Pacific and Indian oceans.

HIPPOCAMPUS

Phylum	-	Chordata
Group	-	Vertebrata
Sub-phylum	-	Gnathostomata
Series	-	Pisces
Class	-	Osteichthyes
Sub-class	-	Actinopterygii
Order	-	Syngathiformes
Genus	-	*Hippocampus*

1. It is commonly known as '*Sea-horse*', because its anterior end is shaped like the neck and head of a horse.
2. Body is divisible into head, trunk and tail, measuring 5-17 cm. Tail is prehensile.
3. Head is at right angle to the trunk and looks like that of horse. Mouth is found at the end of snout.
4. The entire body is invested by closely fitting, bony shields, which are produced into tubercles, spines or filaments of varying size at certain places.
5. Gill-clefts are reduced to small openings and covered by opercula.
6. Dorsal fin is single; ventral and caudal fins are absent.
7. Pectoral fins are small and transparent found on either side of head.
8. Female has a small anal fin. Male contains a *brood pouch* which carry eggs until they hatch (parental care).
9. It is found in the costal water of Indian and Atlantic Oceans.

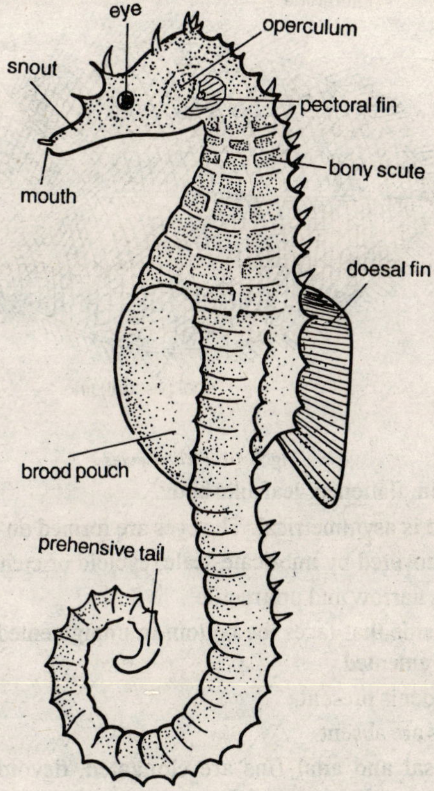

Fig. 3.9. *Hippocampus.*

PLEURONECTES

Phylum	-	Chordata
Group	-	Vertebrata
Sub-phylum	-	Gnathostomata
Series	-	Pisces
Class	-	Osteichthyes
Sub-class	-	Actinopterygii
Order	-	Pleuronecteformes
Genus	-	*Pleuronectes*

1. It is commonly known as '*flat fish*'.
2. It is marine, bottom dweller fish measuring about 50 cm in length.

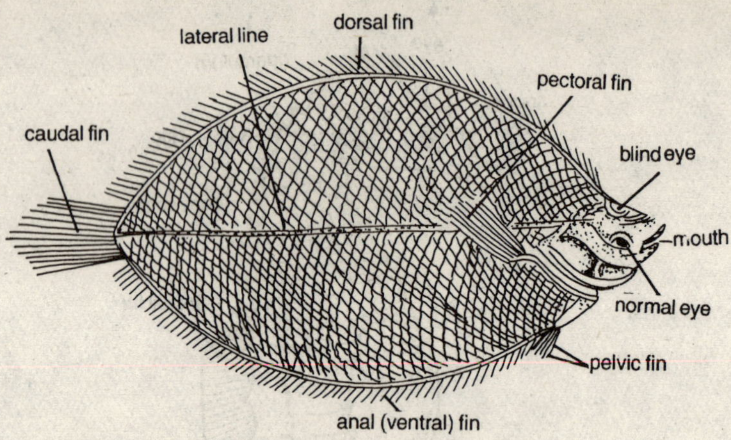

Fig. 3.10. *Pleuronectes.*

3. It has thin, flattened, leaf-like body.
4. The head is asymmetrical. The eyes are formed on the left side.
5. Body is covered by imbricate scale, cycloid or ctenoid.
6. Mouth is narrow and protractile.
7. The left side that faces the bottom in unpigmented while the right side is pigmented.
8. Air-bladder is present.
9. Spiracles are absent.
10. The dorsal and anal fins are elongated, devoid of spines and continuous with the caudal fin.
11. Pectoral fins are placed hight up.
12. *Oviparous.*
13. It feeds on molluscs.
14. It is found at all the tropical and temperate coasts.

TETRADON

Phylum	-	Chordata
Group	-	Vertebrata
Sub-phylum	-	Gnathostomata
Series	-	Pisces
Class	-	Osteichthyes
Sub-class	-	Actinopterygii
Order	-	Tetrodontiformes
Genus	-	*Tetradon*

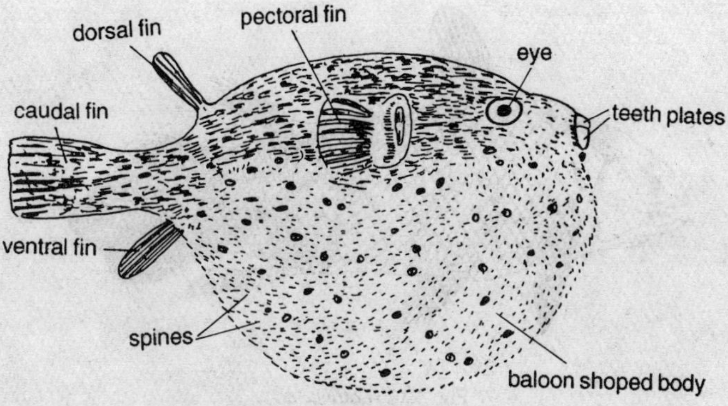

Fig. 3.11. *Tetradon.*

1. It is commonly known as '*globe fish*'.
2. Body is light brown along the back and dark brown bands are found from the back upto the sides.
3. Head and anterior part of body are very large and without scales.
4. Mouth is bounded by lips and armed with dental plates, each split by a median suture.
5. Eyes are large and somewhat protruding.
6. Pectoral fins are large and lateral. Pelvic fins are developed. Caudal fin is broad and entire.
7. Spinous dorsal and ventral fins are absent.
8. Abdomen contains dilatable scas.
9. Air bladder is horse-shore-shaped.
10. It contains a powerful alkaloid poison called *tetradotoxin* thus eating this fish may be fatal.
11. It is marine fish and found in the Tropical, Sub-tropical, Atlantic and Indian Seas. Several found in brakish water and fresh water.

LATIMERIA

Phylum	-	Chordata
Group	-	Vertebrata
Sub-phylum	-	Gnathostomata
Series	-	Pisces
Class	-	Osteichthyes
Sub-class	-	Crossopterygii
Order	-	Coelacanthiformes
Genus	-	*Latimeria*

Fig. 3.12. *Latimeria.*

1. It is a steel-blue fusiform fish, measuring about 1.5 meter long.
2. There are two pairs of external nares.
3. Fins are lobed and highly mobile. The anterior dorsal fin is not lobed. Median fins are separate.
4. Tail is trilobed and diphycercal.
5. Air bladder is reduced and acts as hydrostatic organ.
6. Internal nares are absent.
7. Spiracles are present.
8. Intestine has spiral valve.
9. Endoskeleton has much cartilage.
10. Urinary bladder is paired.
11. It is found in the East Coast of South Africa.

PROTOPTERUS

Phylum	-	Chordata
Group	-	Vertebrata
Sub-phylum	-	Gnathostomata
Series	-	Pisces
Class	-	Dipnoi
Order	-	Lepidosireniformes
Genus	-	*Protopterus*

1. It is commonly known as '*African lung fish*', adapted for burrowing life.
2. The body is elongated and eel-like and is completely covered by cycloid scales.
3. Head contains small eyes, nostrils and mouth.

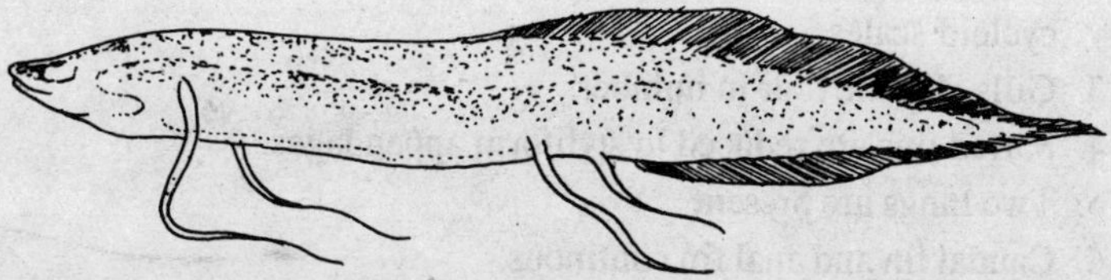

Fig. 3.13. *Protopterus.*

4. The pelvic fins and pectoral fins are reduced to slender appendages.
5. There are 6 branchial arches and 5 clefts.
6. Two lungs extending through out the body cavity.
7. Lateral line system is well developed.
8. The larval gills retained as vestigial organ throughout life. There are four pairs of external gills in larva.
9. It comes to surface to engulf air.
10. It undergoes *aestivation* during summer.
11. It is found in the swamps of great African Continent, the Nile, Congo basin etc.

LEPIDOSIREN

Phylum	-	Chordata
Group	-	Vertebrata
Sub-phylum	-	Gnathostomata
Series	-	Pisces
Class	-	Dipnoi
Order	-	Lepidosireniformes
Genus	-	*Lepidosiren*

Fig. 3.14. *Lepidosiren.*

1. It is commonly known as '*South Americal lung fish*', lives in the burrow.

2. Body elongated and eel-like, one meter in length, and covered by cycloid scales.
3. Gills-slits are four in number.
4. Paired fins are reduced to styliform appendages.
5. Two lungs are present.
6. Caudal fin and anal fin continous.
7. Lateral line system is present.
8. During breeding season, male has a vascular filament on the pelvic fin. It is considered to be a respiratory organ.
9. Parental care is shown by forming 'L' shaped nests. Larva possess four gills.
10. Undergoes *aestivation* during summer.
11. It is found in Amazon and Paraguay basins and plains of South Africa.

NEOCERATODUS

Phylum	-	Chordata
Group	-	Vertebrata
Sub-phylum	-	Gnathostomata
Series	-	Pisces
Class	-	Dipnoi
Order	-	Ceratodiformes
Genus	-	*Neoceratodus*

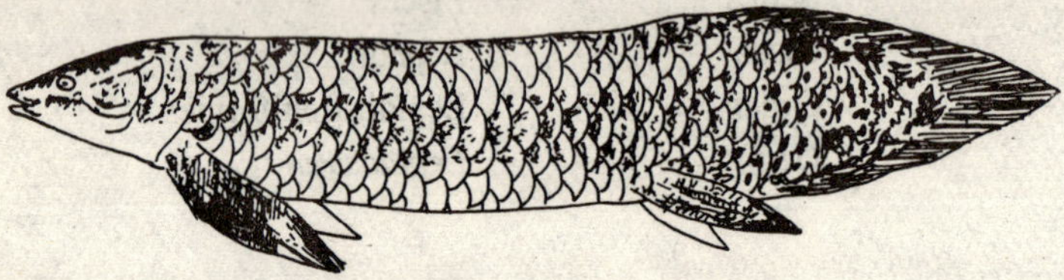

Fig. 3.15. *Neoceratodus.*

1. It is commonly known as '*Australian lung fish*'.
2. Body is elongated and measures 1-2 meters in length. Body covered with large cycloid scales.
3. Gill-slits five pairs.
4. Paired fins placed low, pectoral fins near the head and pelvic fins near the tail.
5. Tail is diphycercal.

6. Single air bladder (lung) opens into oesophagus by a long duct.
7. Male does not process vascular filament on pelvic fin.
8. The paired fins are modified into paddles.
9. Larva without external gills.
10. Internal nares are present.
11. It does not undergo aestivation during summer.
12. It is found in Buret and Mary rivers of Australia and Queensland.

Revision Questions

1. Give the characters and classification of Chondrichthyes upto orders giving examples.
2. GIve characters and classification of Osteichthyes upto orders giving suitable examples.
3. Write short notes on : (a) *Exocoetus*, (b) *Torpedo*, (c) *Hippocampus*, (d) *Amia*, (e) *Pristis*, (f) Dipnoi.

4

Scoliodon

Phylum	-	Chordata
Group	-	Vertebrata
Sub-phylum	-	Gnathostomata
Series	-	Pisces
Class	-	Chondrichthyes
Sub-class	-	Selachii
Order	-	Pleurotremata
Genus	-	*Scoliodon*

Distribution

Genus *Scoliodon* is widely distributed. It occurs along the coasts of Indian Ocean, Arabian Sea, Bay of Bengal, Sri Lanka, Philippines, West Indies etc. There are nine species of *Scoliodon*. In India there are four species *S. sorrakowah*, *S.palasorrah*, *S.walbeehmi* and *S.dumerilii*. The most common species is *S.sorrakowah* which means 'black shark' in Tamil.

Habits and Habitat

Like others sharks, *Scoliodon* is also a marine shark that lives in open seas. It is a carnivorous fish feeds on other fishes, crabs and other crustaceans. It is a voracious feeder and can smell from a long distance. The shape of the body is modified for fast swimming. Sexes are separate, sexual dimorphism is well marked. Fertilization is internal and it gives birth to young-ones i.e. *viviparous.*

EXTERNAL FEATURES

The body is long, laterally compressed, spindle shaped tapering at both the ends. The dorsal body surface is dark gray and the ventral side is pale white. A full specimen measure about 60 cm in length.

The body is divisible into three parts: The *head*, *trunk* and *tail* but there is no distinct boundries between them.

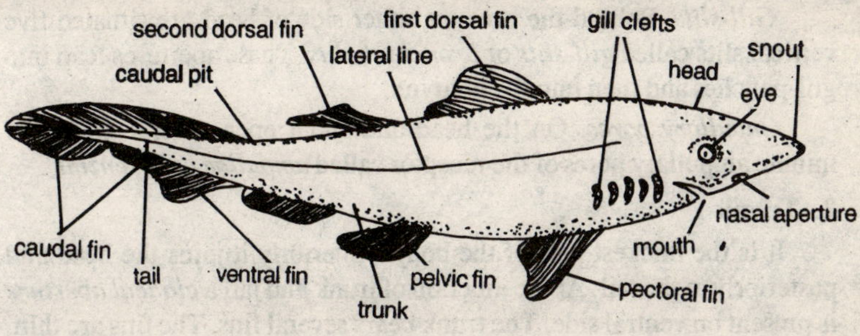

Fig. 4.1. *Scoliodon.*

1. Head

The head is dorso - ventrally flattened and produced forwards to form the snout. The head bears mouth, nares, external gill - slits and ampullary pores.

Mouth. A crescentic mouth is situated on the ventral surface of the head. It is bounded by the *upper* and *lower jaws.* The skin is folded over both the jaws so as to form the *lips* which are immovable. Each jaws in provided with one or two rows of teeth.

Nares. Two crescentic apertures, the *nares*, are present on the ventral surface of the snout anterior to mouth. They lead into the olfactory sacs. They are olfactory in function because the internal nares are absent so they can not function as respiratory organ.

Eyes. Two large circular eyes are situated at the sides of the head. Each eye in provided with *upper* and *lower eyelids* which are merely the outgrowths of skin and are movable. A *nictitating membrane* lies along the antero-ventral border of each eye and can be drawn over the eye during danger. The pupil is narrow and vertical.

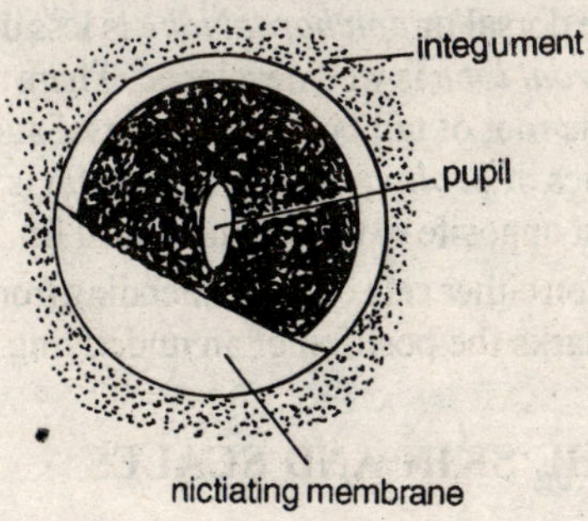

Fig. 4.2. *Scoliodon.*Eye.

Gill-slits. Behind the eyes on either side of head are situated five vertical slits called *gill-slits* or *branchial-slits*. These apertures lead into gill-pouches and then into the pharynx.

Ampullary pores. On the head and snout open several groups of minute ampullary pores of the receptor called *ampullae of Lorenzini'*.

2. Trunk

It is the thickest part of the body. Anteriorly it joins the head and posteriorly to the tail. At the junction of trunk and tail a *cloacal aperture* is present on ventral side. The trunk bears several fins. The fins are thin, flat outgrowths of skin with muscles and are supported by cartilaginous rods, the *fin-rays*. There are two sets of fins: unpaired median fins and paired lateral fins. There is a single median fin, the *first dorsal fin*. It lies on the dorsal side and is triangular in outline. There are two pairs of paired or lateral fins: *pectoral fins* and *pelvic fins*. The pectoral fins are situated anteriorly, one on the either side of trunk behind the gill-slits. The pelvic fins are much smaller and are situated on the ventral side at the junction of trunk and tail enclosing cloacal aperture. In the male, each pelvic fin bears a stiff rod-like grooved copulatory organ, along the inner edge, known as *clasper*. The *abdominal pores*, situated on small papillae, are present on either side of *cloacal aperture*. There pores open into coelomic cavity.

3. Tail

It is the posterior most part of the body behind the trunk and narrows down posteriorly. In the posterior region, it waves upwards at a small angle. This type of tail is called *heterocercal*. Tail bears three median or unpaired fins : *second dorsal fin*, *caudal fin* and *anal* or *ventral fin*. The second dorsal so called posterior dorsal is triangular but much smaller lying between first dorsal and caudal fins. The tail fin extends along the dorsal and ventral surface of the tail in the middle line and forms a dorsal and a ventral lobe. The dorsal or *epichordal lobe* is less developed while the ventral or *hypochordal lobe* is well developed. There are shallow pits called *caudal pits* at the root of tail both on the dorsal and ventral sides. These are characteristics of *Scoliodon*. The ventral fin is situated on the mid-ventral line almost opposite to the second dorsal fin.

A faint lateral line on either side of body extending from the tail, trunk and head regions. It marks the position of an underlying sensory lateral line canal system.

THE SKIN AND SCALES

Skin

The skin forms external covering of the body and performs a number of functions. Skin is composed of two layers: the epidermis and the dermis or corium.

Epidermis. It is ectodermal in origin and consists of several layers of flattened cells. The innermost layer has cuboidal cells and is known as *Malpighian layer* or *stratum germinativum*. These cells rest on a *basement membrane*. Cells of outer layer become progressively flattened and keratinized towards the surface but the stratum corneum, present in terrestrial vertebrates, is not present in fishes. Among the epidermal cells numerous unicellular mucous glands are present which secrete mucous. The mucous makes the body slippery.

Dermis. The dermis lies beneath the epidermis and is composed of fibrous connective tissue with collagen fibres and various types of cells. The dermis is differentiated into two regions: outer layer containing few loose fibres form *stratum laxum*. In the stratum laxum the pigment cells are present. These cells are branched and contain pigment granules. The inner layers with compact fibres form *stratum compactum*. Blood capillaries, nerve fibres and muscle fibres are present in dermis.

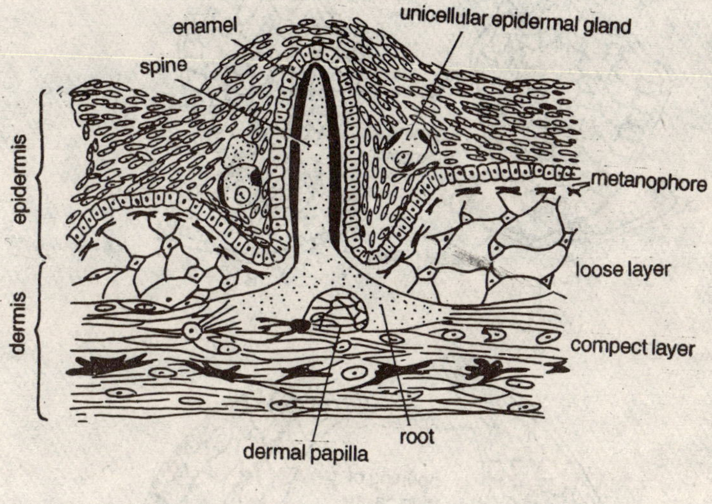

Fig. 4.3. *Scoliodon*. V.S. of skin.

Functions

1. It forms the external covering of the body and protects the body against injury and infections.

2. Secretion of mucous glands makes body slippery and protects the animal from its enemies.
3. The skin and mucous secreted by its glands help the fish in regulating, to some extent, the osmotic exchange of water and ions between the body fluids and the surrounding medium.
4. The derivatives of the skin plays an important part in the metabolic activities of the body and also provide protection to the body.
5. Chromatophores of various kinds, present in the dermis, give beautiful colour patterns to the body.
6. It communicates to animal, changes in the surroundings by means of receptors.

Scales

In most of the fishes, the skin is covered with an exoskeleton in the form of scales. The cat-fishes are devoid of any scales. In *Scoliodon, odontoid* or *placoid scales* are present. These are minute dermal denticles which are arranged in regular oblique rows and form the exoskeleton of shark.

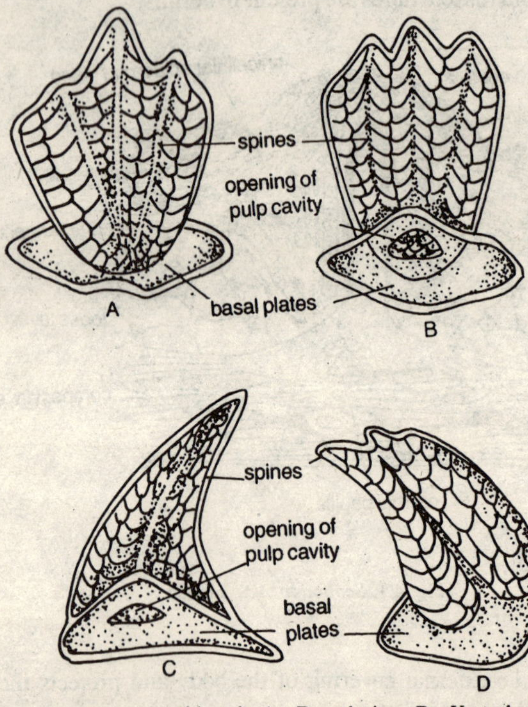

Fig. 4.4. *Scoliodon.* Placoid scale A—Dorsal view, B—Ventral view, C—Lateral view, D—Dosso-lateral view.

Each placoid scale consists of two parts : a diamond shaped *basal plate* and a flat *trident spine*. The basal plate embedded in the skin and firmly held by Sharpey's fibres and connective tissue fibres. The centre of the basal plate is perforated by an aperture known as *opening of pulp cavity* through which blood vessels and nerve from dermis supplied to the scale. The basal plate is formed of a trabecular calcified tissue closely allied to the cementum of mammalian tooth.

The trident spine projecting out of the skin is backwardly direction and is formed by one median and two lateral spines. In structure, a placoid scale resembles a tooth. The spine has an external covering of enamal like, hard, transparent material called *vitrodentine*. It is followed by a layer of *dentine* encasing a *pulp cavity* from which several branching dentine *canaliculi* radiate in different directions. Pulp consists of odontoblasts, blood vessels and nerve fibres.

Development of a Placoid Scale. Some part of scale is formed by epidermis and some by dermis. During its development, the mesodrmal cells aggregate beneath the epidermis. These cells divide and grow due to which the epidermis is pushed upwards and a small papilla is formed. The

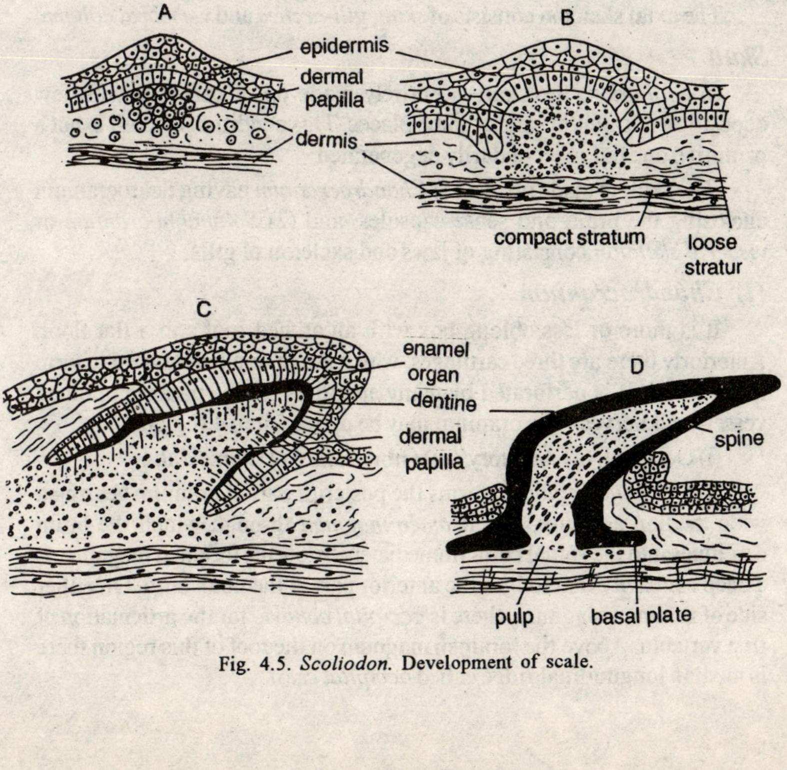

Fig. 4.5. *Scoliodon*. Development of scale.

cells of the Malpighian layer of this region become columnar and glandular forming the *odontoblasts* which form dentine. The epidermal cells above the dentine forms the enamel forming organs which is made up of columnar *ameloblast cells*. The dermal papilla assumes the shape of a basal plate and a spine.

The central cells of the papilla do not calcify and form the pulp. The spine of the scale is gradually pushed towards the-surface. The mesenchymal cells of the dermis, surrounding the dermal papilla, secrete a substance like the cement of teeth to form the basal plate of scale. Blood capillaries and nerve fibres extend into the central cavity through the aperture at the lower part of the basal plate and provides nourishment to the developing scale.

SKELETON

In *Scoliodon*, the endoskeleton is made up of cartilage only and provides system of rigid levers to which muscles are attached. It can be divided into:

I. Axial Skeleton. II Appendicular Skeleton.

I. Axial Skeleton

The axial skeleton consists of *skull*, *gill-arches* and *vertebral column*.

Skull

In *Scoliodon*, the skull is entirely made up of cartilage with few deposition of calcium salts at some places. This condition of skull is not a primitive one but is secondarily degenerated.

The skull is made up of a (1) *Chondrocranium* having neurocranium enclosing the brain and sense capsules, and (2) *Splanchnocranium* or *visceral skeleton* consisting of jaws and skeleton of gills.

(1) Chondrocranium

It is more or less oblong box with an arched roof and a flat floor. Anteriorly there are three cartilages, which are prolonged forward to form the rostrum. It is perforated by many apertures for the passage of blood vessels and nerves. The cranium may be divided into 4 regions:

(i) Occipital, (ii) Auditory, (iii) Orbital and (iv) Ethmoidal.

(i) *Occipital region*. It forms the posterior part of cranium bearing a wide median aperture, the *foramen magnum* through which the brain continues into the spinal cord. Immediately below the foramen magnum, is a deep concavity that lodges the anterior part of the notochord. On either side of foramen magnum, there is *occipital condyle* for the articulation of first vertebra. Above the foramen magnum on the roof of this region there is median longitudinal ridge called *occipital crest*.

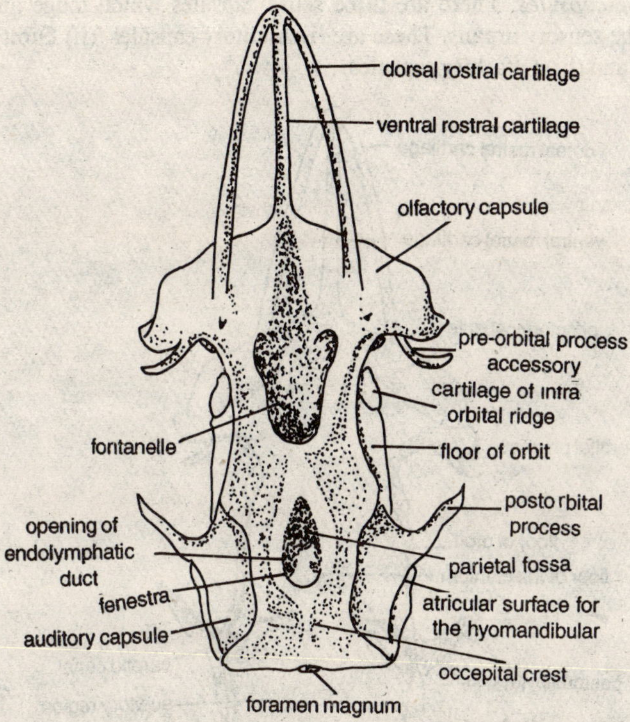

Fig. 4.6. *Scoliodon.* Skull, (Dorsal view).

(ii) *Auditory region.* It has a medium oval depression, the *parietal fossa*, on the dorsal surface. It bears two pairs of apertures in the posterior half. The anterior apertures are known as *openings of endolymphatic ducts* and are smaller. The posterior apertures are larger and known as *openings of perilymphatic spaces.* The ventral surface of auditory region bears a pair of grooves, the *carotid canals* which contain the openings of external and internal carotid arteries.

(iii) *The orbital region.* It has a large roughly oval caity on the dorsal surface, known as *anterior fontanelle.* It is covered by a membrane of connective tissue.

(iv) *Ethmoidal region.* It gives off, from its base, a slender bar known as *ventro-median rostral cartilage* that extends forward and upward to meet a pair of dorso - lateral rostral cartilages which are arising from the roof of olfactory capsules.

Sense-capsules. There are three sense capsules which lodge and protect the sensory organs. These are (i) Auditory capsules, (ii) Orbital capsules and (iii) Olfactory capsules.

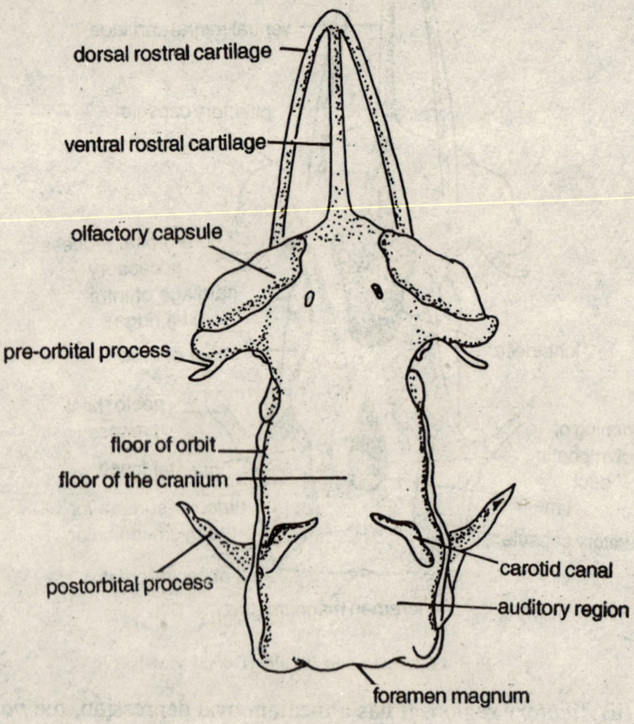

Fig. 4.7. *Scoliodon.* Skull (Ventral view).

(i) *Auditory capsules.* They lie on either sides of the posterior part of cranium. They enclose the membranous labyrinth. Each capsule is bounded by three ridges: anterio-dorsal, posterio-dorsal and ventral. These are the markings of anterior vertical, posterior vertical and horizontal semicircular canals. A *postorbital groove* lies beneath the ventral ridge.

(ii) *Orbital capsules.* The orbits lodge the eyeballs and their muscles. The dorsal boundary of each orbit is formed by a *supra orbital ridge*. Anteriorly it is bounded by short slender *preorbital process* running backward and outward from the olfactory capsule. The posterior boundary is formed by longer and stouter *postorbital process*. It is ventrally bounded by lateral expansion of *infraorbital ridge*. The infraorbital ridge bears in a small *antorbital process* to which the upper jaw is attached by ligament. A large *optic foramen* lies in the middle of orbit for the passage of optic nerve.

(iii) *Olfactory capsules*. They lodge and protect the olfactory sacs. They are oval in shape and having thin walls. Both the capsules are separated from each other by their vertical cartilage, the *internasal septum*. Each capsule bears an opening which leads into cranial cavity. Through this aperture the olfactory lobe of the brain meets the olfactory sac of its side. A short, stout *ethmopalatine ridge* extends backward from each capsule to meet the upper jaw.

(2) Splanchnocranium

It is made up of a series of visceral arches which develop in the pharyngeal wall between the visceral clefts. They encircle the buccal cavity and the pharynx. In *Scoliodon* there are seven visceral arches:

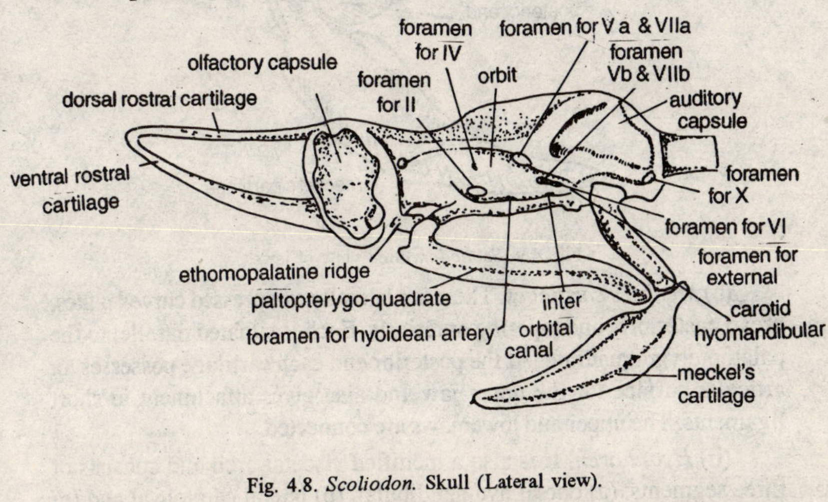

Fig. 4.8. *Scoliodon*. Skull (Lateral view).

(i) A mandibular arch, (ii) A hyoid arch and (iii) Five branchial arches.

(i). *Mandibular arch*. It is especially modified and is a complete ring. It forms the upper and lower jaws, possessing teeth. Each half of this arch consists of two elements (a) an upper *palatopterygo-quadrate cartilage* and *(b)* a lower *Meckel's cartilages*

(a) *Palatopterygo-quadrate cartilage*. Each palatopterygo-quadrate is a strong, curved rod which is compressed posteriorly. It has a deep groove along the anterior two-third of its ventral border. Near the anterior end it gives off a small *palatine* or *orbital process* to which ethmo-palatine ligament is attached. The ligament connects the upper jaw with the cranium.

The palatopterygo-quadrate cartilage of two sides are joined together infront in the median line to form the upper jaw. The posterior ends fit into the articular surface of the two Meckel's cartilages. The mucous membrane lining of the grooves of the jaw bears several rows of teeth.

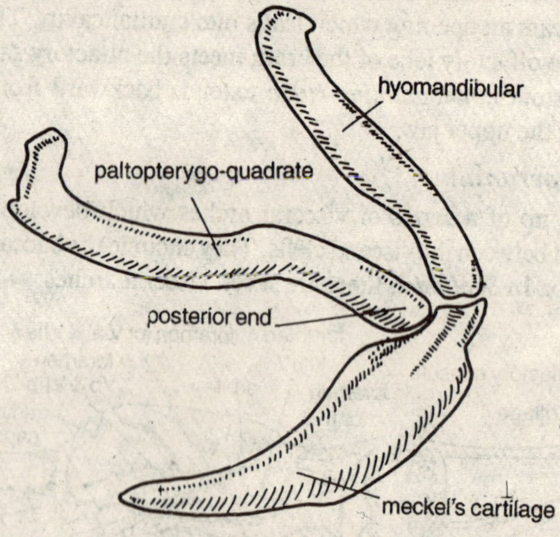

Fig. 4.9. *Scoliodon*. Side view of jaws.

(b) *Meckel's cartilages*. These are laterally compressed curved plates, broad posteriorly and tapering anteriorly. Each is situated parallel to the palatopterygo-quadrate. At the posterior end each cartilage possesses an articular surface for the upper jaw and also gives attachment to short ligaments. The upper and lower jaws are connected.

(ii) *Hyoid arch*. It is also a modified visceral arch and consists of three segments: (a) dorsal hyomandibular, (b) lateral ceratohyal and (c) ventral basihyal.

(a) *The hyomandibular*. It is a short, stout rod that articulates closely with the sides of cranium just below the postorbital groove and ventrally with the *Meckel's cartilage* as well as *basihyal*. Thus the upper and lower jaws are not directly articulated with the cranium but are only joined with it through the hyomandibular cartilage, therefore, it is known as *suspensorium*. This type of jaws suspension is known as *hyostylic*.

(b) *The ceratohyal*. It is slightly, longer than the hyomandibular and is situated in the floor of the buccal cavity, behind the Meckel's cartilage, joining with the median basihyal at its lower end.

The hyomandibular and ceratohyal support the anterior wall of the first gill-slit and bear slender, branched branchial rays or gill-rays along their posterior margin.

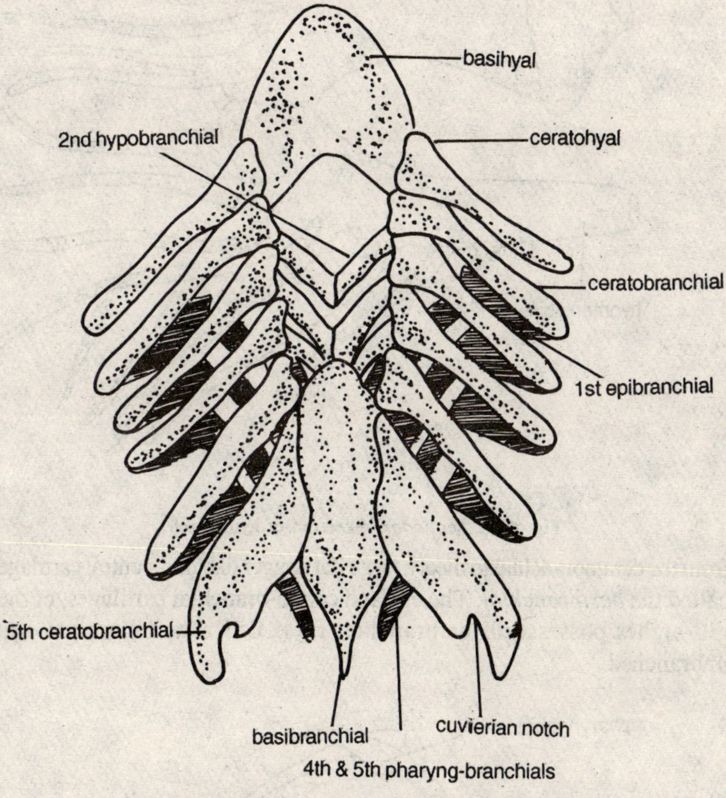

Fig. 4.10. *Scoliodon*. Ventral view of visceral skeleton.

(c) *The basihyal*. It is a broad median plate situated in and supporting the buccal floor. It is rounded anteriorly and is continued behind in two horns which are attached to the ceratohyals.

(iii) *Branchial arches*. Each arch consists of four segments: a long flat dorsal *pharyngo-branchial* running forward and outward from its free median end; a rod-like dorso-lateral *epibranchial* running backwards and out-wards from the pharyngo-branchial; a rod-like ventro-lateral *ceratobranchial* extending forwards and inwards from the epibranchial; and a short rod-like ventral *hypobranchial* running backwards and inwards

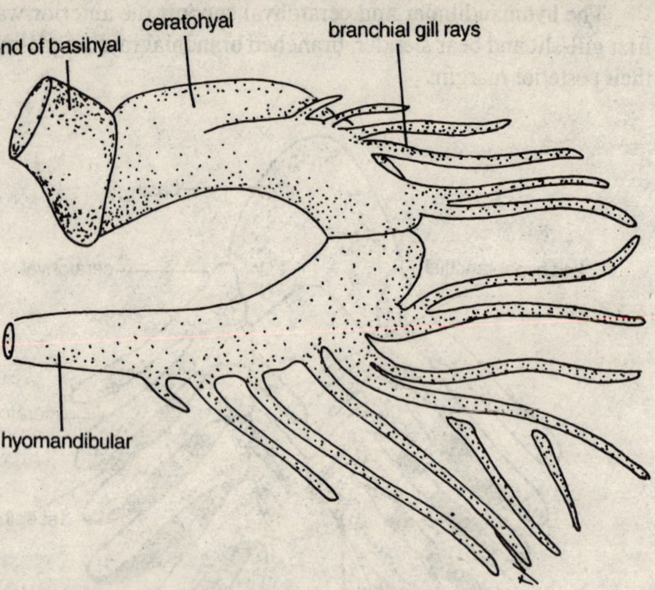

Fig. 4.11. *Scoliodon.* Skeleton of hyoid arch.

from the ceratobranchial to meet a large flat dagger like mid-ventral cartilage called the *basibranchial*. The epi-and cerato-branchial cartilages, of the gill-arches possess gill or branchial rays. Gill rays are simple and unbranched.

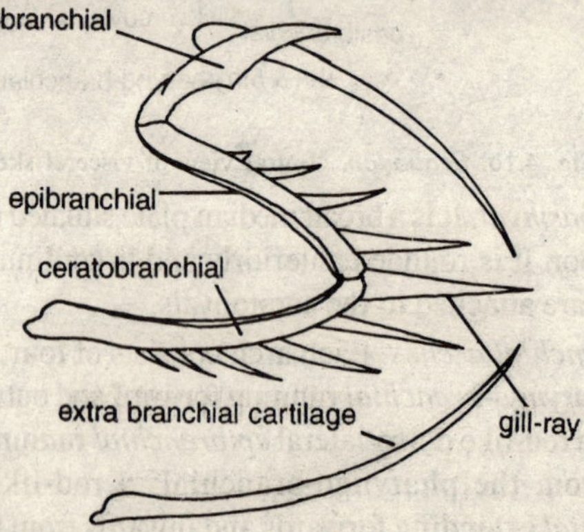

Fig. 4.12. *Scoliodon.* Part of skeleton of first branchial arch.

Other branchial arches show individual variations.

Vertebral column

The vertebral column of *Scoliodon* comprises about 130 vertebrae. These are cartilaginous, solid structures pierced by the *notochord.* The vertebrae are of two types: (a) *Trunk vertebrae* and (b) *Tail vertebrae.*

(a) Trunk vertebra

It is roughly ring-like in appearance. The typical body of the vertebra consists of a ventral thick, cylindrical *centrum* which is *amphicoelous,* (concave anteriorly as well as posteriorly). The centrum has a narrow canal at its centre and is strengthened by calcification in four wedge-shaped areas arranged like a cross known as *Maltese cross.* In between the calcified areas there are uncalcified areas. The centrum with a cross-like calcification is known as *astero-spondylous.* The centrum bears dorsally a *neural arch* enclosing a *neural canal.* The neural canal lodges the spinal cord. Either side of the neural arch is made up by a *vertebral neural plate.* This plate tapers dorsally, leaving a gap infront and behind. These gaps are filled by *intervertebral neural plates.* Dorsally the neural arch is completed by two median plate-like *neural spines.* Between the vertebral and intervertebral neural plates occur separate apertures for the exit of dorsal and ventral roots of spinal nerves. To each ventro-laterd border of centrum is attached a broad plate like *transverse process.*

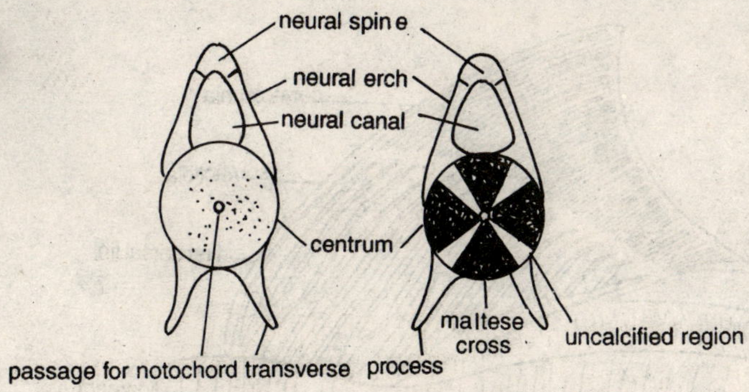

Fig. 4.13. *Scoliodon.* Trunk vertebra A—Anterior view, B—T.S. of Trunk vertebra.

(b) Tail vertebra or Caudal vertebra

The transverse processes meet and fuse to form below centum a *haemal arch,* which encloses a canal called *haemal canal.* Through haemal canal caudal artery and vein pass. The haemal arch in continued into backwardly directed *haemal spine*.

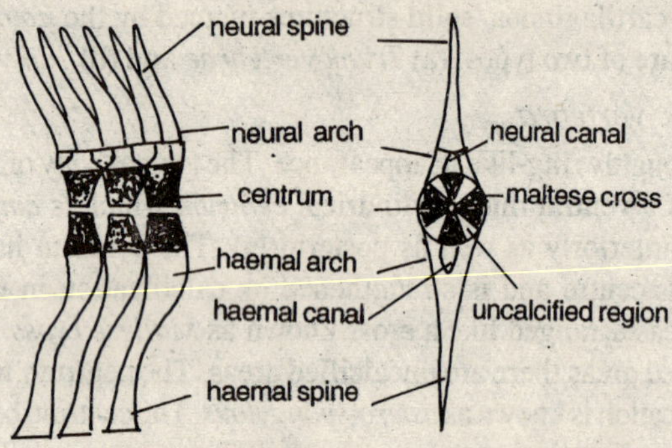

Fig. 4.14. *Scoliodon*. A Caudal vertebra. A—Sagittal section, B—T.S. of vertebra.

The end of the vertebral column is bent upwards and the ventral lobe of the tail-fin is unequal, such an asymmetrical tail is called *heterocercal*.

II. Appendicular Skeleton

The appendicular skeleton consists of the skeleton of median fins, paired fins, pectoral girdle and pelvic girdle.

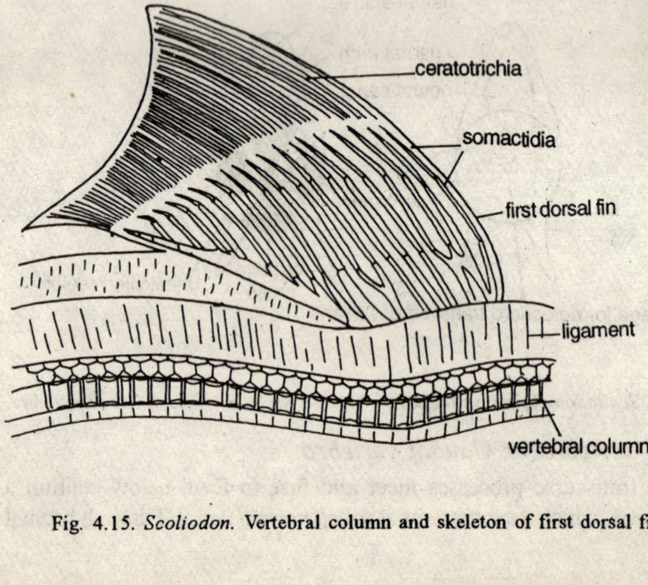

Fig. 4.15. *Scoliodon*. Vertebral column and skeleton of first dorsal fin.

Median fins

The skeleton of median fins consists of series of cartilaginous rods, called *somactidia* or *pterygiophores.* The somactidia possess a double series of horny fin-rays or *ceratotrichia.* Each somactidium is connected with vertebral column by a wide band of ligament. There are 21 somactidia in the anterior or first dorsal fin, but fewer in posterior or second dorsal fin and ventral fin. In the caudal fin somactidia are absent and dorsal and ventral lobes of the fins are supported by the neural and haemal spines of vertebral column.

Pectoral girdle

It is embedded in the lateral and ventral body wall, just posterior to the gills. It is not directly connected with the axial skeleton, being held by ligaments. It is a 'U' shaped stucture made of right and left halves united midventrally. Each half has a thin, flat ventral part, the *coracoid* and a thick rod-like *scapula.* At the junction of the two parts of each half, there is a triple facet, the glenoid surface, for the articulation of the basal cartilages of the pectoral fin. The coracoid bears few foramina for blood vessels.

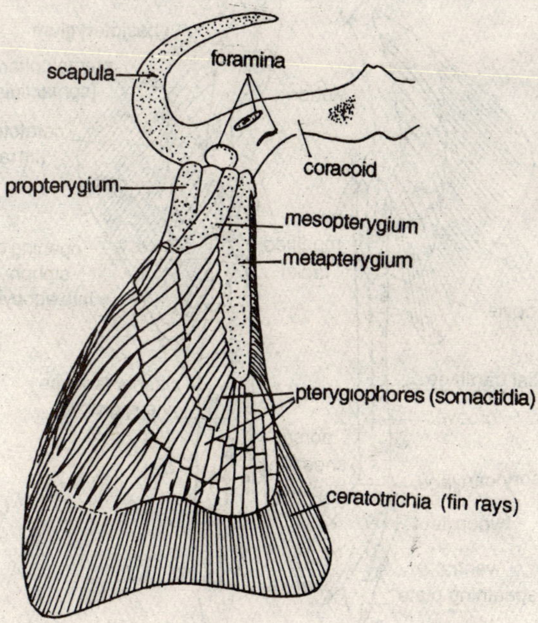

Fig. 4.16. *Scoliodon.* Pectoral girdle and fin.

Pectoral fins

Each articulates with the pectoral girdle and consists of three basal cartilages called, *protopterygium, mesopterygium* and *metapterygium*. The radial cartilages are three segmented rods present at the distal ends of basal cartilages. These are called the *somactidia*. The protopterygium bears two, mesopterygium bears two and metapterygium bears 18 somactidia. The distal pieces are very long, unbranched, close-set rods of elastin in the distal part of the fin they are called *ceretotrichia* or *fin rays*.

Pelvic girdle

It is a flattened cartilaginous bar, the *ischiopubic bar*, which lies transversely infront of cloaca. It is straight in the middle but bent at the ends, which are produced dorsolaterally into short blunt *iliac processes*. Each end is perforated by a nerve foramen and bears a facet the *acetabular surface* for the articulation of basal cartilage of the pelvic fin.

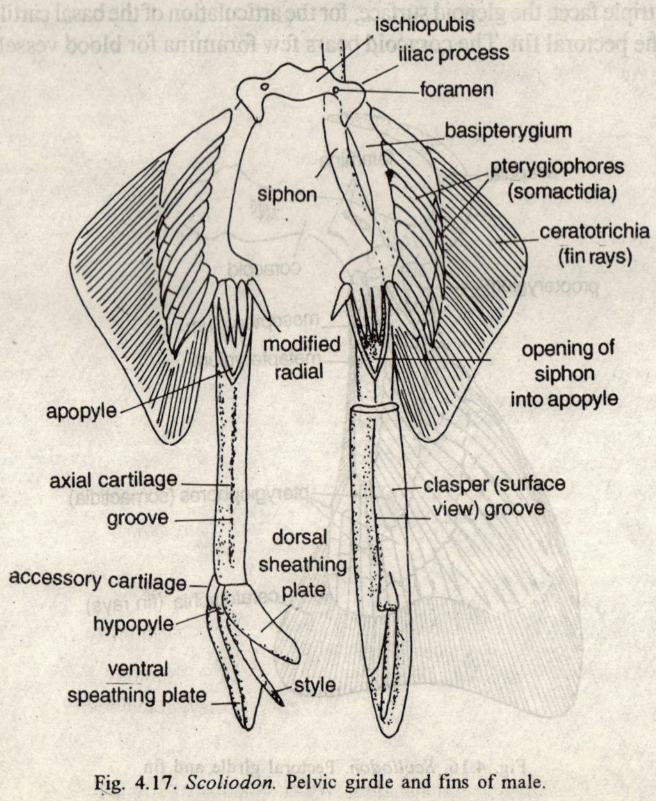

Fig. 4.17. *Scoliodon*. Pelvic girdle and fins of male.

Pelvic fins -

Skeleton of pelvic fin include a single large and curved basal cartilage, the *basipterygium*. 15 or more slender *somactidia* and numerous distal fin-rays or *ceratotrichia* are present.

In male, the two pelvic fins are modified posteriorly to form the *claspers*. Each clasper is attached to the distal end of basipterygium. It has a long, stout, rod-like *axial cartilage*. A groove runs along the dorsal side of this cartilage. The anterior opening of groove or *apopyle* communicates with cloaca and a *siphon*. The posterior opening or *hypopyle*, opens on a sharp pointed *style* enclosed within a dorsal and a ventral sheathing place. At the upper end of style, there is also a serrated *accessory cartilage*.

BODY CAVITY OR COELOM

The body cavity of *Scoliodon* is a large space divided into two very unequal compartments:

(a) *Pericardial* and (b) *Perivisceral coelom* or *abdominal coelom*. Both these cavities are separated by the *septum transversum*, which is a partition of tough connective tissue. The pericardial cavity is a small, triangular space lying beneath the pharynx and surrounding the heart. It is lined by a double layered *pericardium*. It is filled with a clear and colourless pericardial fluid. The pericardial cavity opens into the abdominal cavity through *pericardio-peritoneal aperture*.

The abdominal cavity is very large and surrounds the viscera and opens to outside through a pair of abdominal pores situated on the papillae, one on either side of the cloacal aperture. It is also filled with colourless

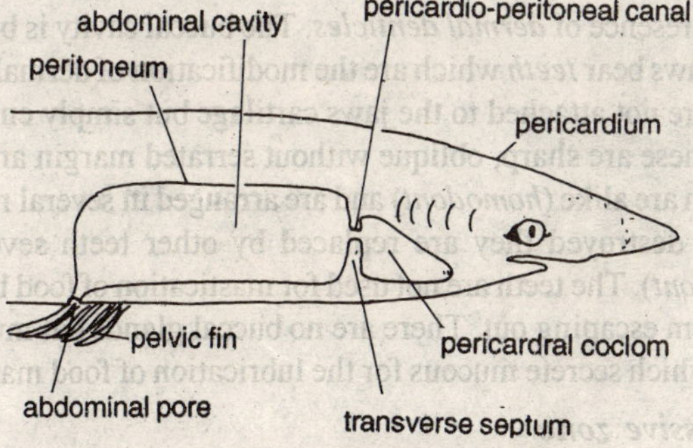

Fig. 4.18. *Scoliodon*. Coelomic body cavities.

coelomic fluid. It is lined by peritoneum. The peritoneum also covers the surface of visceral organs. Beneath the vertebral column, there is a double fold of peritoneum, the *mesentery* by which the alimentary canal suspended in the coelom. Different parts of the alimentary canal are bound to one another by incomplete mesenteries called *omenta*. The *gastro-hepatic omentum* attaches the liver with stomach. The *gastrosplenic omentum* attaches the spleen with the stomach.

DIGESTIVE SYSTEM

The digestive system includes the *alimentary canal* and *digestive glands*.

Alimentary Canal

The alimentary canal of *Scoliodon* starts from mouth and terminates into anus. It is differentiated into four zones or regions.

1. The *Ingressive zone* includes the mouth and buccal cavity
2. The *Progressive zone* which includes the pharynx, oesophagus and stomach.
3. The *Degressive zone* includes the small intestine.
4. The *Egressive zone* includes the large intestine.

1. The ingressive zone

Mouth. The mouth is a wide crescentic opening on the ventral side of the head. It is bounded by the folds of skin which are sometimes called the upper and lower lips. Mouth leads into the buccal cavity.

Buccal cavity. It is a spacious, dorso-ventrally flattened chamber. It is lined with thick mucous membrane. On the floor of buccal cavity lies the *tongue*, which is a thick, flat, non-muscular, non-glandular and non-protrusible fold of mucous membrane. The mucous membrane is rough due to the presence of *dermal denticles*. The buccal cavity is bordered by *jaws*. The jaws bear *teeth* which are the modification of dermal denticles. The teeth are not attached to the jaws cartilage but simply embedded in the skin. These are sharp, oblique without serrated margin and smooth. All the teeth are alike (*homodont*) and are arranged in several rows if they are lost or destroyed they are replaced by other teeth several times (*polyphyodont*). The teeth are not used for mastication of food but prevent the prey from escaping out. There are no buccal glands but mucous cells are found which secrete mucous for the lubrication of food materials.

2. Progressive zone

Pharynx. The buccal cavity posteriorly leads into the pharynx. On either side of pharynx an oval pit of spiracle and five separate vertical

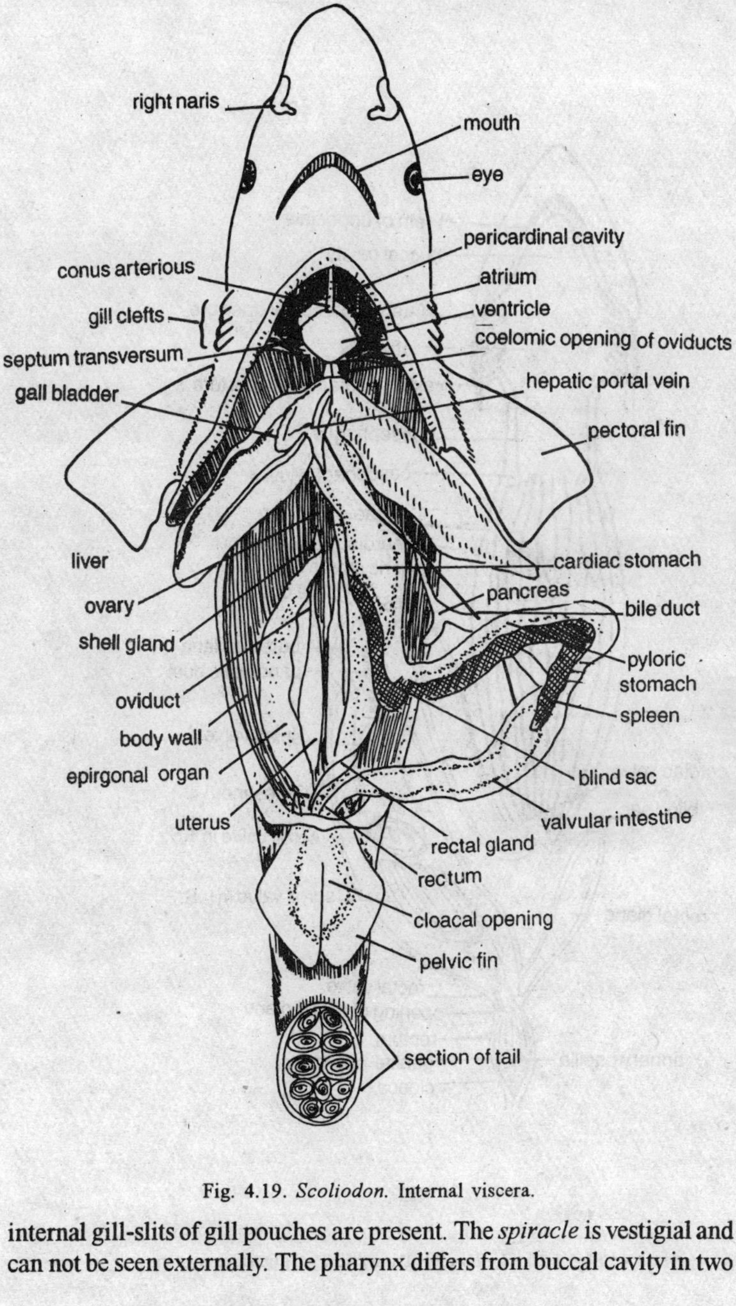

Fig. 4.19. *Scoliodon.* Internal viscera.

internal gill-slits of gill pouches are present. The *spiracle* is vestigial and can not be seen externally. The pharynx differs from buccal cavity in two

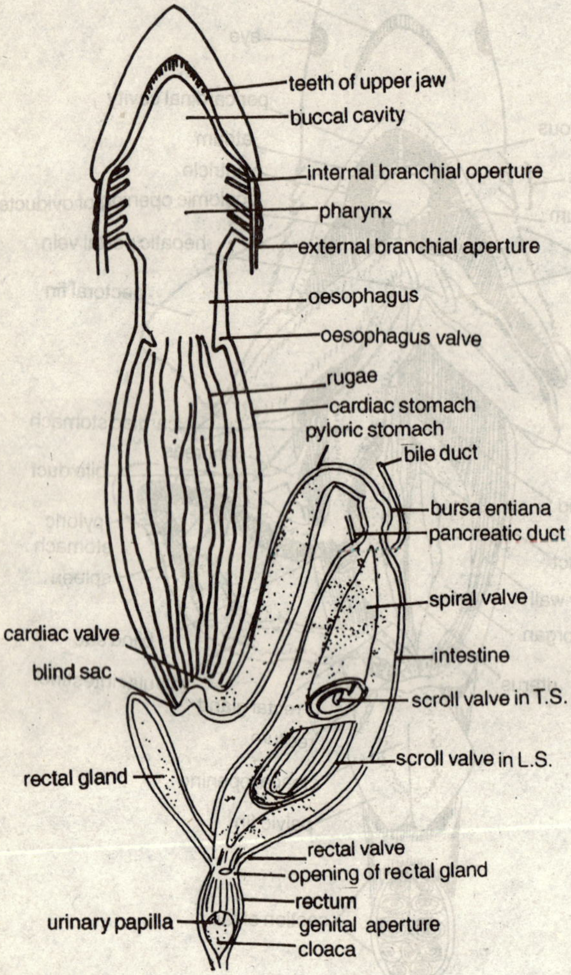

Fig. 4.20. *Scoliodon*. Dissection of digestive system.

respects (i) its inner lining is made up of endoderm, and (ii) presence of gill-slits and spiracles.

The mucous lining of pharynx also possess dermal denticles and mucous cells.

Oesophagus. The pharynx narrows down posteriorly so as to form the oesophagus. The oesophagus is a short tube having thick muscular wall. The mucous membrane is raised into longitudinal folds known as *rugae*. Thus the cavity of oesophagus is completely closed, preventing the entering of water which continuously passes through pharynx for respiration.

Stomach. The oesophagus opens behind into the stomach. At the junction of oesophagus and stomach lies a cresentic *oesophageal valve*. The stomach is a 'J' shaped structure having two limbs, the proximal limb is longer and broader and is known as *cardiac stomach* where as the distal limb is short and narrow known as *pyloric stomach*. At the junction of cardiac and pyloric stomachs, their is a small growth called *blind sac*. The longitudinal folds of the cardiac stomach are continued anteriorly into the longitudinal folds of oesophagus and posteriorly they end in blind sac. At the junction of cardiac and pyloric stomaches there is a *cardiac valve*. The lining of pyloric stomach is smooth and only a few folds are present near its distal end. At the end of pyloric stomach, there is a circular band of muscle fibres, called *pyloric valve* which guards the opening of stomach into a small but thick walled muscular chamber called *bursa entiana,* which is continuous into intestine. The stomach is suspended in body cavity by a double fold of peritonium called *mesogaster.*

3. Degressive zone

Intestine. Stomach leads into intestine. It is a straight wide tube about the diameter of cardiac stomach in the middle region. Anteriorly it receives the bile duct from liver and pancreatic duct from pancreas. Thus the anterior part may be called *duodenum* while remaining part as *ileum*. The internal surface of intestine is increased by the presence of a characteristic fold of mucous membrane known as *scroll valve*. One edge of this valve is attached to the intestinal mucosa, while the other edge is rolled up longitudinally on itself into a scroll, having an anti-clock wise spiral of about two and a half turns. This scroll valve not only increases the absorptive surface of the intestine but slows down the passage of food through ileum.

4. Egresseive zone

Rectum. The intestine opens into the rectum. The rectum is a short and narrow tube opening behind into the cloaca through the anus. The

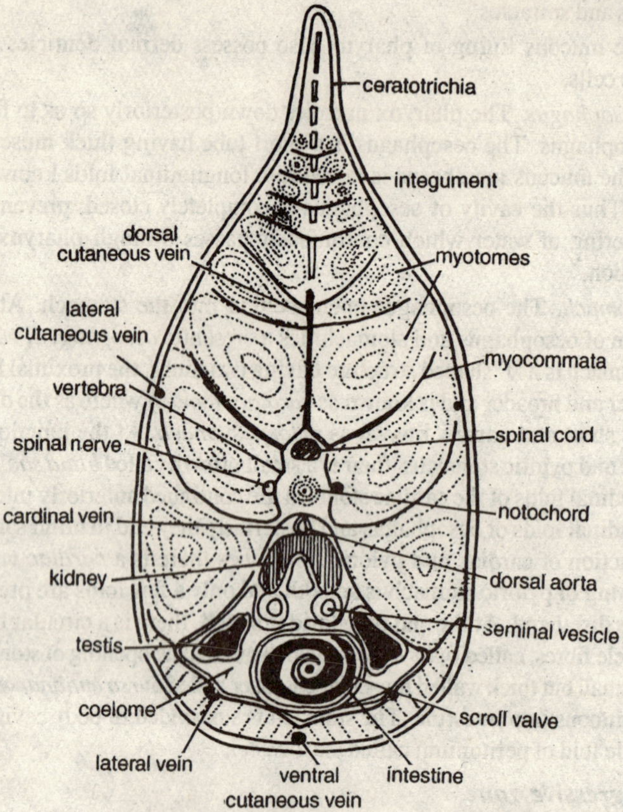

Fig. 4.21. *Scoliodon.* T.S. of the body of male through-intestine.

tubular *rectal* or *caecal gland* opens closely into the rectum. The function of rectal gland is unknown. The cloaca also receives the openings of urinogenital ducts. A *rectal valve* is present at the junction of intestine and rectum.

Cloaca. It is a short terminal chamber which receives the alimentary canal and urinogenital ducts. It also receives a pair of abdominal pores leading from the splanchnocoel.

Digestive Glands

The glands which are associated with alimentary canal are *liver*, *pancreas*, *gastric glands*, *intestinal glands*, *rectal gland* and *spleen*.

Liver. Liver of *Scoliodon* is a large bilobed yellow or mustard-coloured gland occupying a considerable length and space of the abdominal cavity. The two lobes i.e. right and left lobes are united below the oesophagus by a short, strong, double fold of peritoneum called *falciform ligament*. A 'V' shaped *gall bladder* storing bile over the right lobe of liver is present. The *bile duct* or *ductus choledochus* carries bile into the intestine and opens into the dorsal wall. Bile duct also receives branches from lobes of liver. The liver secretes bile, stores glycogen and fat and destroys worn out erythrocytes of blood. The fat constitutes 80% of the shark.liver. Very large fatty liver in shank has hydrostatic function.

Pancreas. The pancreas is a compact, whitish gland pressed between the two limbs of stomach. It consists of two lobes: a longer dorsal lobe running parallel to the posterior part of cardiac stomach, and smaller ventral lobe lying close to the pyloric stomach. A pancreatic duct runs through the whole length of the gland and opens into the intestine, opposite the aperture of bile-duct. Pancreas secretes pancreatic juice for digestion.

Gastric glands. These are microscopic to sub-microscopic occurs in the wall of stomach. They secrete gastric juice for digestion of food.

Intestinal glands. These glands are also lie in the wall of intestine. They secrete *succus entericus*.

Rectal gland. A small finger-like structure is attached to the dorsal wall of rectum. It is highly vascular and composed of lymphoid tissue. Its function is not well known.

Spleen. It is a large gland attached to cardiac stomach. It is not a digestive gland but a lymphoid organ which produces lymphocytes.

Food, feeding and digestion

Scoliodon is a carnivorous predator feeds on other fishes, crustaceans, molluscs and annelid worms. It is a *voracious feeder* and can smell food from a long distance. The food is swallowed without mastication. The mouth is opened by *coracomandibular muscle* and is quickly closed by *adductor mandibular muscle*. There is no digestion in buccal cavity, pharynx and oesophagus. Actual digestion starts in cardiac stomach. The gastric juice contains pepsin and hydrochloric acid. Pepsin converts proteins into peptones and proteoses. In the intestine, the bile intestinal and pancreatic juices are mixed with semidigested food. Complete digestion occurs here and reabsorption takes place in intestine. The scroll valve provides large surface area for absorption and it also retards the quick passage of food. The undigested food is passed to the rectum from where it is stored for sometimes in cloaca and then eliminated out through the cloacal aperture.

RESPIRATORY SYSTEM

Scoliodon is an aquatic animal thus it depends on the oxygen dissolved in water for its respiratory requirements. The dissolved oxygen is taken with the help of special respiratory organs known as *gills*. The respiratory system compries: the *respiratory tract* and the *respiratory organs.*

The respiratopy tract

It consists of mouth, buccal cavity and pharynx. Ventrally situated mouth opens into the buccal cavity which posteriorly continues as the pharynx. The water is convey to gills through this tract.

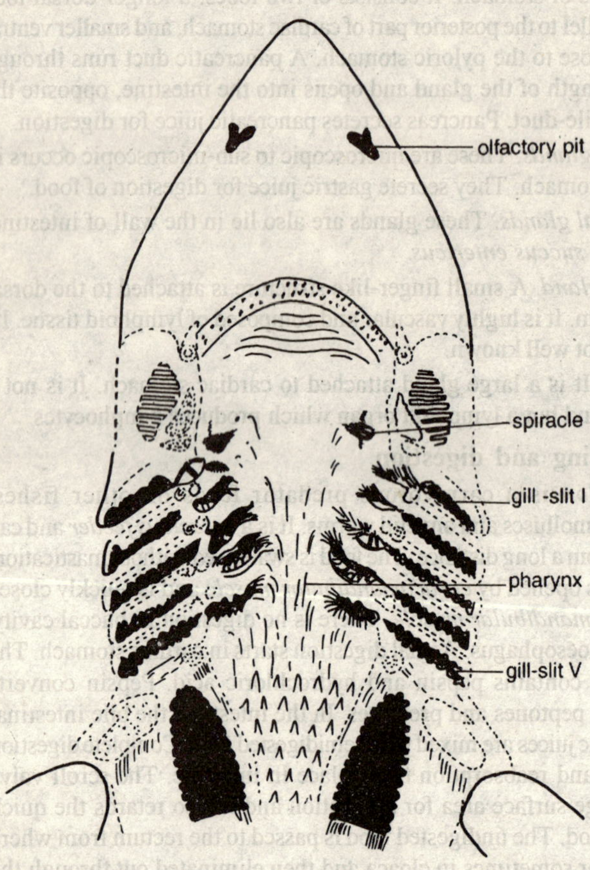

Fig. 4.22. *Scoliodon.* Pharynx exposed to show gill-pouches.

The respiratory organs

The respiratory organs consist of five pairs of *gill-pouches* containing *gills*. The gill-pouches are situated in lateral walls of the pharynx and are arranged in a series on either side. Each gill- pouch communicates on one side, with the pharynx by a large *internal branchial aperture* and with the outside by a narrow *external branchial aperture*. Two adjacent gill-pouches are completely separated from each other by a vertical *interbranchial septum*. The septa are composed of fibro-muscular tissue containing blood vessels and nerves. The inner or pharyngeal border of each septum is supported by a cartilaginous *gill-arch* or *branchial arch* with its slender *branchial ray* passes outwards to support the rest of septum. The endodermal mucous membrane of the gill-pouch is raised into a series of horizontal leaf-like folds called *gill-lamellae* or *gill-filaments*. These constitute the gill-proper for exchange of gases as they are richly supplied with blood capillaries.

Each septum has two sets of gill lamellae, one on the anterior and other on the posterior face. When an interbranchial septum bears two sets of lamellae, it is known as *holobranch* or *complete gill*. A single set of branchial lamellae is regarded as a *hemibranch* or *demibranch* or *half gill*. A gill-pouch, thus, contains the posterior demibranch of one gill and the anterior demibranch of succeeding gill. In *Scoliodon* there is one demibranch borne by the posterior face of the hyoid arch, four holobranchs borne by four branchial arches; and the fifth arch is gill-less or *abranch*.

Each inter-branchial septum is prolonged externally beyond the branchial lamellae into a thin branchial flap and contains two sets of muscles, *vertical* and *horizontal*.

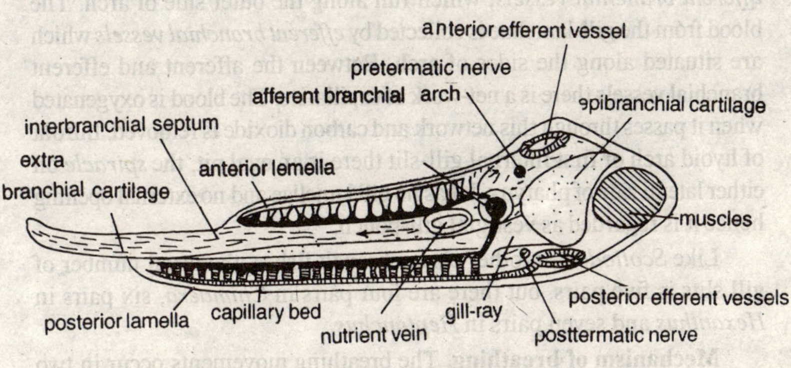

Fig. 4.23. *Scoliodon.* Section of holobranch.

The gills of *Scoliodon* are *lamelliform* in which the *gill-lamellae* are attached to the inter-branchial septum all along their length. In bony fishes the gills are *filiform* as the interbranchial septum is much reduced.

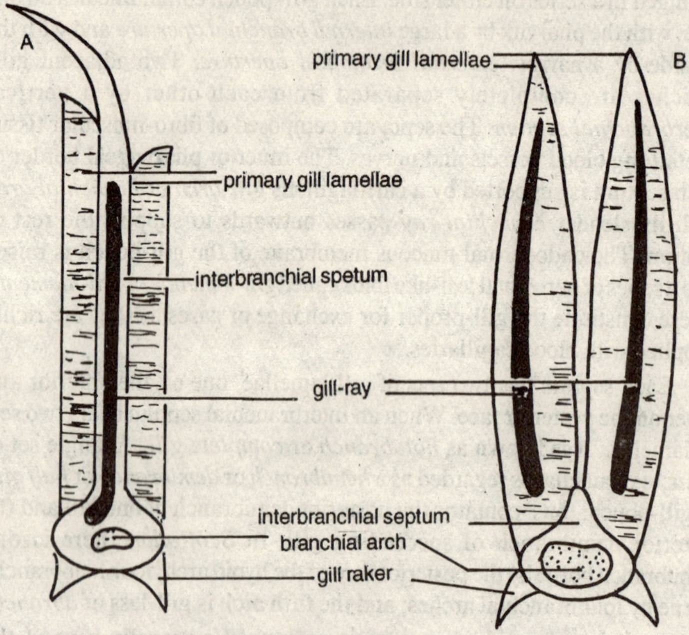

Fig. 4.24. Types of gills, A—Lamelliform, B—Filiform.

Each gill-lamella possess many sinusoides. They receive blood from *afferent branchial vessels*, which run along the outer side of arch. The blood from the gill-lamellae is collected by *efferent branchial vessels* which are situated along the sides of arch. Between the afferent and efferent branchial vessels there is a net-work of capillaries. The blood is oxygenated when it passes through this network and carbon dioxide is removed. Infront of hyoid arch or first internal gill-slit there is an oval pit, the *spiracle* on either lateral side of pharynx. It has no gill lamellae and no external opening hence it is regarded as vestigial gill-pouch.

Like *Scoliodon*, in other elasmobranch fishes, the usual number of gill slits is five pairs, but there are four pairs in *Chimaera*, six pairs in *Hexanthus* and seven pairs in *Heptanchus.*

Mechanism of breathing. The breathing movements occur in two steps:

1. Inspiration

During inspiration, the *Scoliodon* pulls the lower jaw voluntarily by the contraction of coracobranchial muscles thus the mouth is opened. At the same time the gill-arches expand the wall of pharynx so that the sea-water immediately rushes into the buccal cavity. The external gill-slits are tightly closed by the branchial flaps so as to prevent the entry of water from outside into the pharynx through these slits.

2. Expiration

The mouth is closed by the contraction of *superficial constrictor muscles* and by the *intrinsic constuctor muscles* of the gill-pouches, which raise the lower jaw tightly against the upper jaw, forcing the water into pharynx. Gill arches de-arch with the help of their intrinsic muscles, thus a pressure is exerted on water, which escapes out through the gill-slits. During entire process the oesophagus remains closed by its sphincter muscle.

Physiology of respiration

As mentioned above that the gill-lamellae are richly supplied with blood sinusoids which receive venous blood from afferent branchial

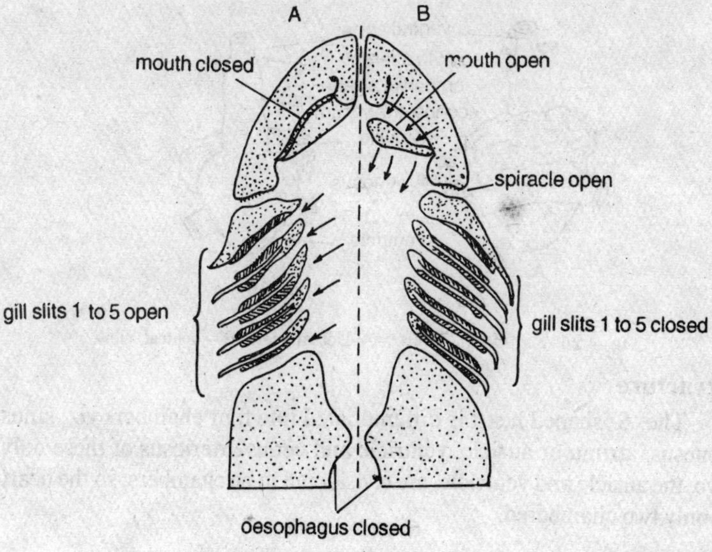

Fig. 4.25. *Scoliodon*. Breathing mechanism. A—Expiration, B—Inspiration.

vessels and pass it to the efferent branchial vessels. The walls of blood capillaries are very thin and permeable. As the gill-lamellae are constantly bathed with fresh water, the oxygen passes into the blood by endosmosis and the carbon dioxide diffused out into the water from blood. Thus the oxygen is conveyed to all parts of body by the blood and the carbon dioxide is brought to the gills is also by blood.

CIRCULATORY SYSTEM

The blood vascular system comprises, the *heart*, *blood vessels*, *blood* and *lymphatic system*.

Heart

The heart is a reddish, conical organ situated in the mid-ventral line beneath the pharynx. It is supported here by the coracoid cartilage of pectoral girdle. It is enclosed in a double layered, transparent membraneous sac, the *pericardium*. The inner layer is the *visceral pericardium* and the outer is *parietal pericardium*. Between the two layers is the *pericardial cavity* which is filled with *pericardial fluid*. The fluid protects the heart from external shocks and provides free movement during contractions. The pericardial cavity communicates with abdominal cavity by paired *pericardio-peritoneal canals* perforating the septum transversum.

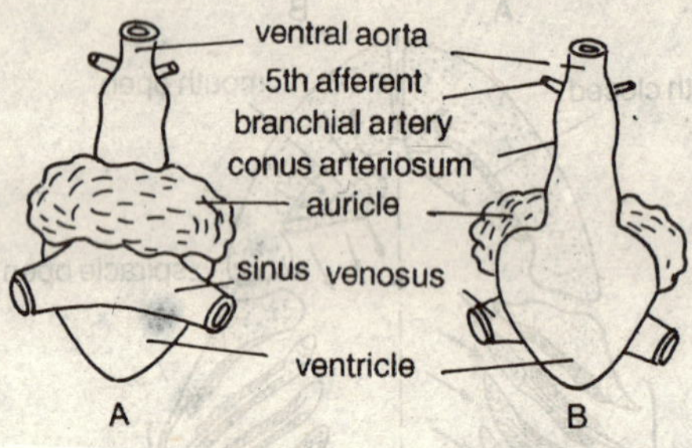

Fig. 4.26. *Scoliodon*. Heart, A—Dorsal view, B—Ventral view.

Structure

The 'S' shaped heart is differentiated into four chambers viz. sinus venosus, atrium or auricle, ventricle and conus arteriosus of these only two, the auricle and ventricle, are considered to be chambers, so the heart is only two chambered.

Sinus venosus

It is somewhat triangular, elastic, thin walled chamber lies on the dorsal side of the ventricle, along the base of pericardial cavity. Laterally it receives two large veins, the *ductus Cuvieri*, one on each side. A pair of *hepatic sinuses* open into it in the posterio-median line. Anteriorly it opens into the atrium by an aperture known as *sinuatrial aperture* guarded by a pair of membranous *sinuatrial valves*. These valves prevent backward flow of blood.

Atrium

It is a large triangular spongy and thin walled sac situated infront of the sinus venosus. Its two lateral sides projecting beyond the ventricle appear like ears (*auricular appendages*). Ventrally it opens into the ventricle by *auriculo-ventricular aperture* guarded by two lip-like valves called *auriculo-ventricular valves*. *Chordae tendineae* attach the lips of this valve with the inner surface of ventricle.

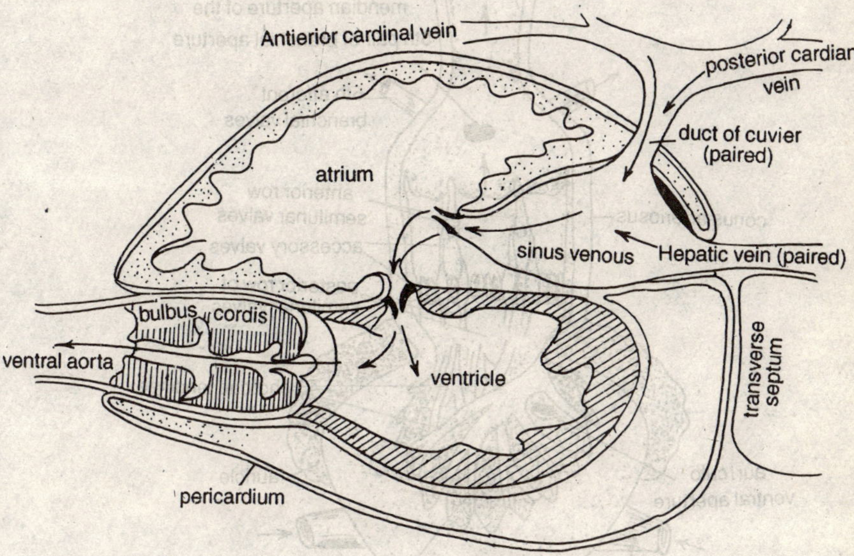

Fig. 4.27. *Scoliodon*. V.S. of Heart.

Ventricle

The ventricle is prominent conical chamber with thick muscular walls. Numerous strends are present in its inner surface due to which it appears spongy in structure. It tapers anteriorly to join the conus arteriosus.

Conus arteriosus

It is a stout muscular tube extending from the ventricle upto the anterior apex of the pericardial cavity. Cavity of conus arteriosus contains two transverse rows of *semilunar valves*. Each row has three valves-one dorsal and two ventro-lateral in position (thus there are six valves in all). An *accesory valve* is present on either side of dorsal valve. The free ends of all the valves are attached to the wall of the ventricle by means of tendinous threads which hold the valves in position. These valves prevent backward flow of blood. Conus arteriosus continues anteriorly into ventral aorta.

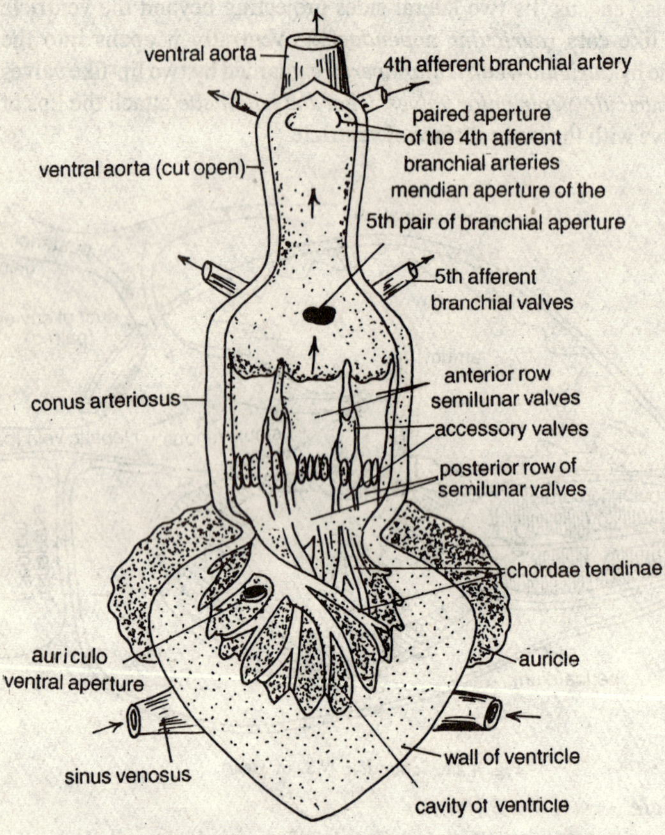

Fig. 4.28. *Scoliodon*. Heart dissected from ventral side.

Working of heart

Heart of *Scoliodon* receives only venous or deoxygenated blood hence it is called as *venous heart* or *branchial heart*. The heart serves to pump the blood. The blood is received from various parts of the body in sinus venosus. Contraction (systole) starts from the sinus venosus forcing the blood into the atrium via sinu-atrial aperture. By the contraction of atrium the blood reaches the ventricle through auriculo-ventricular aperture. In the mean time the sinus-venosus relaxed (diastole). The ventricle contracts and forces the blood into conus, from conus the blood enters the ventral aorta. The deoxygenated blood is pumped to gills for aeration by afferent branchial vessels. After oxygenation, the blood is not sent back to the heart but is forced to various parts of body by the efferent branchial vessels and dorsal aorta. The deoxygenated blood is again returned to the heart from various parts of the body. In this manner during its course of circulation the blood passes through heart only once in its complete circuit. Thus the heart is called *venous heart* and the blood circulation is decribed as *single type* of *circulation*.

The heart beat is initiated by the sinus venous hence may be said as pace-maker.

ARTERIAL SYSTEM

Arterial system supplies blood to all the parts of body. For convenience the arterial system can be studied under five headings:

I. Ventral Aorta and Afferent Branchials

The conus arteriosus continues forward as a single stout artery, the *ventral aorta* that runs beneath the pharynx upto the hyoid arch. The ventral aorta bifurcates near the hyoid arch into right and left *innominate arteries*. Each innominate artery divides into *first* and *second afferent branchial arteries*. The *third, fourth* and *fifth afferent branchial arteries* independently arise from ventral aorta. The fifth branchials of both the sides arise by a common median aperture from ventral aorta.First afferent branchial artery runs along the posterior border of the hyoid arch and supplies the hyoidean hemibranch, similarly, the second afferent branchial artery supplies the first holobranch attached to the third gill arch. The third, fourth and fifth afferent branchial arteries supply, the 2nd, 3rd and 4th holobranchs respectively. The 5th is abranch hence there is no blood supply. The afferent branchials break up into capillaries in the gill-lamellae for aeration of blood.

II. Efferent Branchials and Epibranchials

Oxygenated blood is returned from the gills by nine *efferent branchial arteries* on each side. The first eight efferent branchial arteries unite in pairs to form four complete loops around the first four gill-pouches.

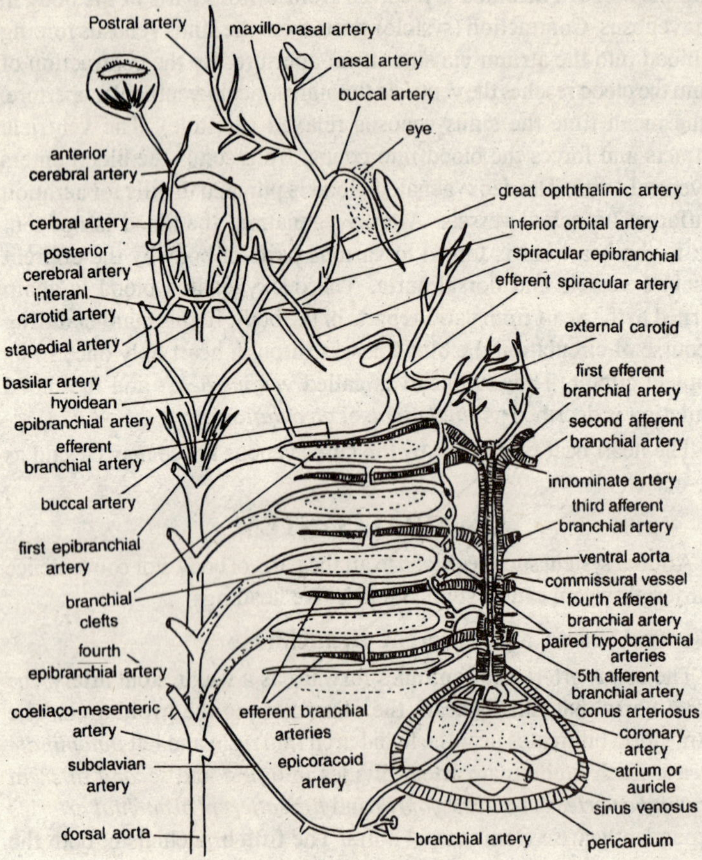

Fig. 4.29. *Scoliodon.* Arterial system.

The single ninth efferent branchial artery runs along the anterior side of fifth gill-pouch. Adjacent loops are interconnected by short *longitudinal connectives*. The efferent branchial arteries are called *pre-trematic* those running along anterior sides, the *post-trematic* run along the posterior sides. Each of four loops give inward an *epibranchial artery*. Thus the four epibranchials run inward and backward and join their fellows of other side mid-dorsally to form the dorsal aorta.

III. Hypobranchial Plexus

It occurs in the ventral wall of pharynx and consists of four thin longitudinal arteries: a pair of *median hypobranchial* and a pair of *lateral hypobranchial arteries*. The lateral hypobranchial arteries are connected externally with the efferent loops by a number of short fine vessels and internally by four fine vessels to the median hypobranchial arteries. Two median hypobranchial arteries are connected by transverse vessels and posteriorly they unite to form a single artery known as *median coracoid artery*. The coracoid artery sends a single branch to heart known as *coronary artery* and a *pericardial artery* to the pericardium. The pericardial artery gives off an *epicoracoid artery* that joins the sub-clavians.

IV. Arteries of Head

The head region is supplied by three pairs of arteries arising from first pair of pre-trematic or first efferent. These are:

1. External carotid artery

It arises at the antero-ventral corner of the first collector loop and travels along the outer surface of the hyoid arch. It branches into a ventral *mandibular* that runs forward and outward, gives off small branches to the coraco-mandibular muscles along the inner surface of the lower jaw to supply its muscles; and *superficial hyoid* supplied blood to the, skin muscles and subcutaneous tissue of hyoid arch.

2. The afferent spiracular

It arises from middle of pre-trematic moves straight and curves in spiracle regions. It continues forward as *spiracular epibranchial* and enters the brain box and joins the internal carotid so as to form the *cerebral artery*. Before entering the cranium it gives off the *opthalmic artery* to the eye-ball. The cerebral artery is divided into *anterior* and *posterior cerebral arteries* supplying blood to the brain. The posterior cerebral arteries united to form the *basilar artery* and that runs backward as the *spinal artery*.

3. The hyoidean epibranchial

It arises close to the dorsal end of hyoidean efferent and runs forward and inward to join one *lateral branch* or *radix* of dorsal aorta, thereafter, it gives off a *stapedial* branch and itself continues as the *internal carotid*. The stapedial branch gives an *inferior orbital artery* (supplying several branches to eye-ball) and continues as *superior orbital* to the anterior of the orbit where it gives off a large *buccal artery* to lower jaw, *maxillo-nasal* to upper jaw and nasal sac and *rostral* to rostrum. The internal carotid artery enters the cranium and joins its fellow of other side.

V. The Dorsal Aorta and its branches

The *dorsal aorta* is formed by the union of four pairs of epibranchial arteries. It runs backward along the whole lengh of the body. It gives off the following branches from infront backwards;

1. *Radices*. The anterior end of dorsal aorta bifurcates into two called radices or lateral aortae. Each radix joins with the hyoidean epibranchial of its side.
2. *Vertebral arteries*. One pair to the vertebral column.
3. *Buccal arteries*. There are one pair of fine vessels leaving the anterior end of the dorsal aorta to supply blood to the roof of buccal cavity.
4. *Sub-clavian arteries*. These are paired arteries. Each sub-clavian receives an epicoracoid artery from hypobranchial plexus and gives off a *lateral artery* to the lateral-line organ, *dorso-lateral* to dorsal body wall; *brachial* to the pelvic fin.
5. *Coeliaeo-mesenteric artery*. It is a large unpaired artery, divided into two branches *coeliac* that supplies to oesophagus, cardiac stomach and liver; and *anterior mesenteric* to the intestine, scroll valve and pancreas.
6. *Lieno-gastric artery*. A single artery arises behind the coeliaco-mesenteric artery which supplies to gonad by *genital* artery, to pyloric stomach and spleen by *posterior gastric artery*.
7. *Posterior mesenteric artery*. Median single artery supplies blood to rectal gland.
8. *Iliac arteries*. Paired, supplies blood to pelvic fins.
9. *Parietal arteries*. These are a number of paired arteries that supply to dorsal and lateral body wall, vertebral column, lateral line canals, kidneys and gonads in female.
10. *Caudal artery*. Posteriorly the dorsal aorta continues into tail as caudal artery lying in the haemal canal of caudal vertebrae.

VENOUS SYSTEM

The blood distributed to the various parts of body by the arteries and their branches is collected through capillaries and returned to the heart by the veins. The venous system may be studied under following heads:

1. The Anterior Cardinal System

Two paired sinuses draining the blood from head region constitute this system. These are present on either side:

(1) Anterior cardinal sinus and (2) Inferior jugular sinus.

1. Anterior cardinal sinus

It lies dorsal to the visceral arches. Several small veins return blood from the rostrum and join to form the *anterior facial* or *orbito-nasal vein* leading into the *olfactory sinus*. This sinus passes through a foramen in the floor of the olfactory capsule into the *orbito-nasal sinus* which opens into *orbital sinus*. The orbital sinuses of left and right sides are connected by an *inter-orbital sinus* lying in the inter- orbital canal of the floor of the cranium. Each orbital sinus leads into the *anterior cardinal sinus* through a narrow *postorbital sinus*. The anterior cardinal is a large sinus running backward between the dorsal ends of gill pouches and the muscles of body wall and entering the *ductus Cuvieri* behind. The anterior cardinal also receive blood from the *hyoidean sinus* which extends obliquely from mandibular sympyhysis to the first gill- cleft. Five dorsal nutrient *branchial sinuses* from the gills also join the anterior cardinal sinus by separate openings.

2. *Inferior jugular sinus*

It is located below the gill-pouches and receives blood from the floor of buccal cavity and pharynx. It is formed by the union of *sub-mental sinus* from lower jaw and the *hyoidean sinus*. It also receives five ventral *nutrient sinuses* from the gills. It collects blood from the ventral region of the gill-pouches. Although the dorsal and ventral nutrient sinuses open into the anterior cardinal and inferior jungular sinuses respectively, they are connected within each interbranchial septum. Finally, each inferior jugular sinus opens, on each side, into the *ductus Cuvieri*, which leads into the sinus venosus. Their openings (sinuses) are guarded by semilunar valves.

II. The Posterior Cardinal System or Renal Portal System

It is formed of *postcardinal sinuses*, that running mid-dorsally in the abdominal cavity outside the peritoneum. This system returns blood from kidneys and the dorsal body-wall muscles. Blood is collected from the tail by a median *caudal vein* which traverses the haemal canal of caudal vertebrae beneath the caudal artery. On entering the abominal cavity dorsally to the cloaca, it bifurcates into *right* and *left renal portal veins*. Each renal portal vain continues forward along the dorso-lateral margins of kidney. In the kidney it gives off numerous *afferent renal veins*. Small *parietal veins* from the body-wall join the renal portal veins all along their length. The blood from the sinusoids of the kidney is collected by the *efferent renal veins* which join together to form the *posterior cardinal sinuses*. In the posterior part of kidneys numerous veins join to form a median *inter-renal vein*. Infront of kidneys both the sinuses become enormously enlarged and each receive an *oesophageal vein* from oesophagus and several *genital veins* from a large *genital sinus*. Finally they open into ductus Cuvieri.

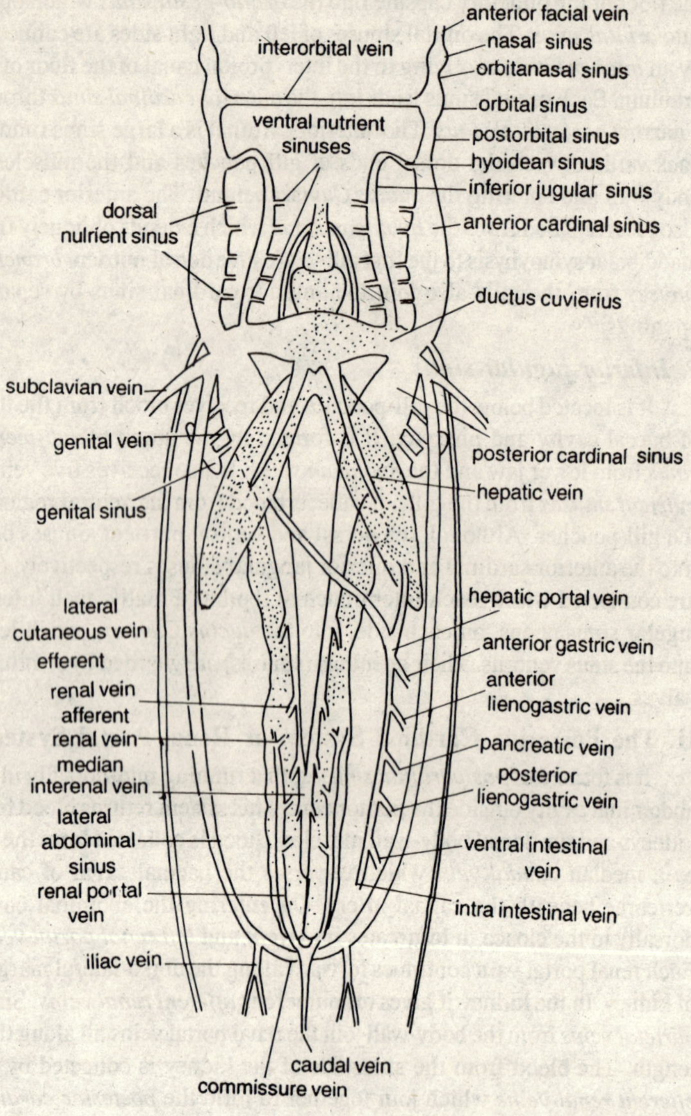

Fig. 4.30. *Scoliodon*. Venous system.

III. The Sub-intestinal or Hepatic Portal System

Blood from the alimentary canal and its associated glands is collected by a number of veins which unite to form the *hepatic portal vein*. The hepatic portal vein receives following veins : *intestinal vein* from scroll valve; *ventral intestinal veins* from ventral wall of intestine; *posterior lieno-gastric vein* from the pyloric stomach and spleen; *pancreatic vein* from pancreas; *anterior- lieno-gastric vein* from anterior part of pyloric stomach and *anterior gastric vein* from cardiac stomach. The hepatic portal vein divides into right and left branches and open into the right and left lobes of liver. Blood from the liver is returned by a pair of large hepatic sinuses. These sinuses open anteriorly into the sinus-venosus by separate apertures opposite the sinu- auricular aperture.

IV. The Lateral Abdominal System

Blood from inner lateral parts of body wall, cloacal region and paired fins on either side is collected by large lateral *abdominal veins*. The two lateral veins are joined posteriorly by a *commissural vein*. Each lateral abdominal receives near its hind end as *iliac vein* which is formed by the union of *cloacal* and *femoral veins*, draining blood from pelvic fins. Anteriorly, each lateral abdominal joins a branchial vein from pectoral fin forming the *sub-clavian vein* which enters the ductus Cuvieri laterally.

V. Cutaneous System

It collects blood from the skin of trunk and tail regions. This system consists of a *dorsal*, a *ventral* and two paired *lateral veins*. They collect blood from their respective regions and pour it into the sinus venosus via ductus Cuvieri.

Blood

The blood consists of a liquid portion, the *plasma* and that contains *corpuscles*. The plasma is colourless contains organic and inorganic substances. The corpuscles are of. two types *red blood corpuscles* or *erythrocytes* and *white blood corpuscles* or *leucocytes*. The erythrocytes are oval, nucleated bodies containing the respiratory pigment, *haemoglobin*. Thus erythrocytes are oxygen carries in the blood and helps in respiration. The leucocytes are capable of amoeboid movement. They help in *phagocytosis*. The blood of elasmobranchs, body-fluid and tissues are characterized by high urea contents about 2 –2.25%. It is probably for osmoregulation in sea-water.

Lymphatic System

The lymphatic system has not been worked out in *Scoliodon*. But in general, there are lymphatic vessels which are superficial as well as deep. The deep vessels follow the course of large veins. All the lymphatics drop into paired longitudinal sinuses which fall into cardinal sinuses. The lymph-nodes are not developed in fishes.

NERVOUS SYSTEM

The nervous system consists of (I) Central nervous system, (II) Peripheral nervous system and (III) Autonomic nervous system.

I. Central Nervous System

The central nervous system including the brain and the spinal cord.

Brain

The brain forms anterior part of the central nervous system. It lies in the brain box or cranium of the skull. It is a soft whitish structure of relatively large in size. The brain in divided into *fore-brain*, *mid-brain* and *hind-brain*.

Fore-brain. It is differentiated into *olfactory lobes*, *cerebrum* and *diencephalon*.

(a) *Olfactory lobes*. From the cerebrum, the *olfactory stalks* arise in anteriolateral angles. They terminate into *olfactory lobes* which are closely applied to the *olfactory sac* of their side. The bulb and tract or stalk containing a cavity or ventricle known as *rhinocoel*. Rhinocoels communicate posteriorly with the paracoels of cerebrum.

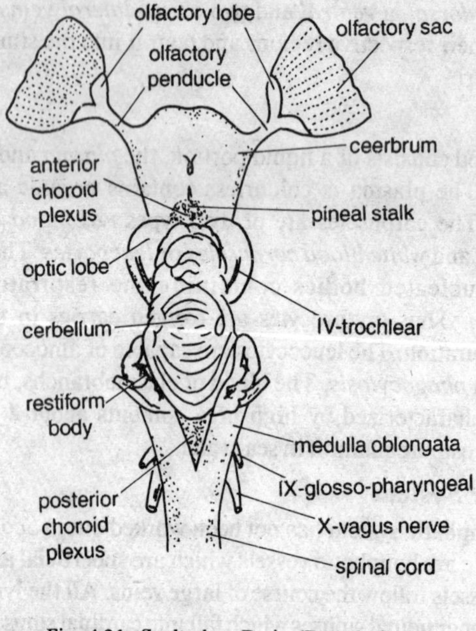

Fig. 4.31. *Scoliodon.* Brain (Dorsal view).

(b) *Cerebrum.* The cerebrum is large, somewhat rectangular mass with smooth surface. There is no median groove to separate the cerebrum into left and right cerebral hemispheres, thus it is an undivided mass. The dorsal surface is smooth but on its mid-ventral surface there is a small opening called *neuropore* through which emerges a pair of *terminal* or *pre-olfactory nerves.* Each nerve bears a ganglion along its course and runs alongside the olfactory tract of its own side to innervate the mucous membrane of olfactory sac. The cerebrum encloses a pair of cavities, the *paracoels* or *lateral ventricles* which are separated by median partition. The paracoels communicate posteriorly with the diacoel by *foramen of Monro.* The roof of each paracoel is thin and called *pallium.* The lateral walls and floor are somewhat thick and known as *corpora striata.*

(c) *Diencephalon.* The cerebrum posteriorly continued into the diencephalon and is completely covered over by cerebellum on the dorsal side. The cavity of diencephalon is called *diacoel.* The roof of which is extremely thin and membranous containing numerous blood vessels forming a folded structure known as *anterior choroid plexus.* The lateral walls are called *thalami.* The floor of diacoel is well developed and is called *hypothalamus.* Hypothalamus sends a large backwardly directed

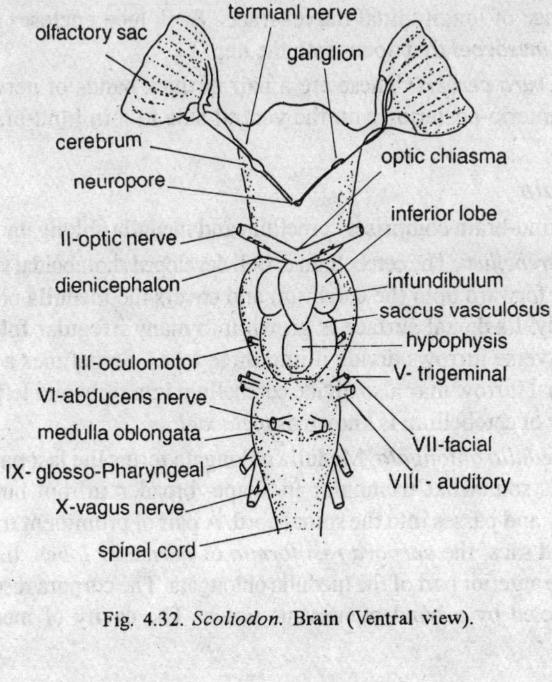

Fig. 4.32. *Scoliodon.* Brain (Ventral view).

down growth, the *infundibulum*. The infundibulum consists of a median large pear shaped lobe and two oval lateral lobes called *inferior lobes*. To the median lobe is attached *hypophysis* forming the *pituitary body*. Large, thin walled vascular sacs, called *sacci vasiculosi* lie on the sides of hypophysis, which act as pressure receptors. Infront of infundibulum lies the *optic chiasma* formed by the crossing of optic nerves. The anterior wall of diacoel is thin and called *lamina terminales*. The roof of diacoel just behind anterior choroid plexus gives rise a long slender process, the *pineal stalk*. Three *commisures*, which are formed of transverse fibres connect the right and left sides of the brain. These are an *anterior commissure* in the lamina terminalis, a *Habenular commissure* infront of pineal stalk and a *posterior commissure* on the dorsal side at the junction of diencephalon with mid-brain. The diacoel communicates with metacoel or 4th ventricle through *iter* or *aqueduct of Sylvius*.

Mid-brain

Mid-brain consists of *optic lobes* and *crura cerebri*.

(a) *Optic lobes*. These are a pair of rounded bodies present on the dorsal side of brain. They are also known as *corpora bigemina*. The optic lobes are almost completely covered over dorsally by cerebellum and ventrally by the infundibular growths. Their lateral sides are enlarged by the presence of longitudinal nerve- tracts. Each lobe encloses an *optic vesicle* or *mesocoel* that opens into the iter.

(b) *Crura cerebri*. These are a pair of thick bands of nerve fibres running anterio-posteriorly on the ventral side to join hind-brain with fore-brain.

Hind-brain

The hind-brain comprises cerbellum and medulla oblongata.

(a) *Cerebellum*. The cerebellum is well developed rhomboidal structure. It extends forward upto the cerebrum and covers the medulla oblongata backwardly. Its dorsal surface is grown into many irregular folds. Two deep transverse furrows divide it into three lobes. Sometimes a median longitudinal furrow may also divide cerebellum into right and left halves. The cavity of cerebellum is known as *metacoel*.

(b) *Medulla oblongata*. Medulla oblongata forms the last part of the brain. It is somewhat triangular in shape, broader infront but tapers posteriorly and passes into the spinal cord. A pair of prominent irregular, thin walled sacs, the *corpora restiformia* or *auricular lobes*, lie on the sides of the anterior part of the medulla oblongata. The corpora restiformia are connected by a band of nervous tissue. The cavity of medulla is known as *myelocoel*. The floor and lateral walls are thick while the roof is thin and highly vascular forming the *posterior choroid plexus*.

The anterior and posterior choroid plexuses secrete a lymph-like fluid, the *cerebrospinal fluid* which is contained in all the cavities of brain.

The brain is covered over by thin, vascular membrane known as *meninx primitiva*. The cranial cavity, is lined by *fibrous connective tissue*, the *endorchis*. There lies a space between two which is filled with cerebrospinal fluid and connective tissue stands.

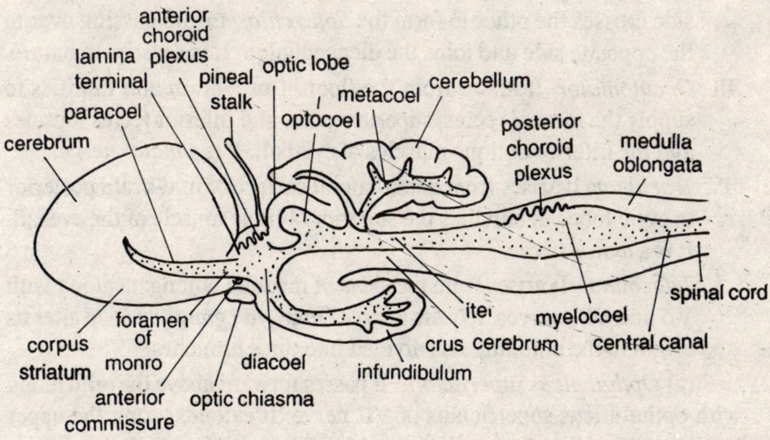

Fig. 4.33. *Scoliodon*. V.S. Brain.

Spinal Cord

The spinal cord extends from medulla oblongata almost to the end of the tail. It has a shallow *dorsal fissure* and a well-marked *ventral fissure*. The spinal cord has at its center a narrow cavity, the *central canal*, which is lined by ciliated epithelium. Anteriorly it opens into the fourth ventricle but is closed behind. In the spinal cord two regions can be distinguished: an outer region, the *white matter*, composed of a non-nervous scaffolding formed by a network of neuroglea cells and medullated nerve fibres; and an internal region of *grey matter* containing chiefly the *neurons*.

II. The Peripheral Nervous System

It includes the cranial nerves and spinal nerves.

Cranial Nerves

There are 10 pairs of cranial nerves in *Scoliodon*. Besides, an additional pair of nerves called 'O' or *terminal nerves* are also present. They are called 'O' because they were discovered long after the others were already numbered.

O or Terminal. This nerve originates through the neuropore present on the ventral side of cerebrum and innervate the olfactory region. It is *sensory* in nature.

I. *Olfactory.* It arises from the olfactory lobe and innervates the mucous membrane of nasal sac. It is *sensory* in nature.

II. *Optic*. The optic nerve arises from the retina of eye. The nerve of each side crosses the other to form the *optic chiasma* and passing over to the opposite side and joins the diencephalon. It is *sensory* in nature.

III. *Occulomotor*. It arises from the floor of mid-brain and ramifies to supply the inernal rectus, superior rectus and inferior rectus muscles and the inferior oblique muscles of eye-ball. It is a *motor nerve*.

IV. *Trochlear*. It arises from dorso-lateral surface of mid-brain posterior to optic lobes. It supplies the superior oblique muscle of the eyeball. It is a *motor* nerve.

V. *Trigeminal*. It arises from the sides of medulla obiongata along with VII and VIII nerves. It bears a large *Gasserian ganglion* soon after its origin in the cranium. It is divided into three branches:

(a) *Opthalmicus superficialis*. It passes forward above the orbit along with opthalmicus superficialis of VII nerve. It extends along the upper border of the orbit and supplies the skin of snout. It is a *sensory* nerve.

(b) *Maxillaries*. It is the main branch of trigeminal and runs along the buccalis branch of VII nerve. It is divided into many fine branches and supplies to the skin of upper jaw, upper lip, lower eyelid and teeth of upper jaw. It is also *sensory* in nature.

(c) *Mandibularis*. It runs along the sides of maxillaries for a short distance and then runs to the lower jaw and supplies to the muscles of lower jaw and tongue. It is a *mixed* nerve.

(d) A fourth branch, the *opthalmicus profundus* comes to be associated with trigeminal nerve secondarily. It supplies to the dorsal skin of snout. It is a *sensory* nerve.

VI. *Abducens*. It is a small nerve arises from the medulla oblongata close to mid-ventral line. It entes the orbit and supplies the posterior rectus muscle of the eye-ball. It is a *motor* nerve.

VII. *Facial*. It.is a large nerve which comes out from the sides of medulla oblongata just beneath the trigeminal nerve. It is a *mixed* nerve and has four main branches:

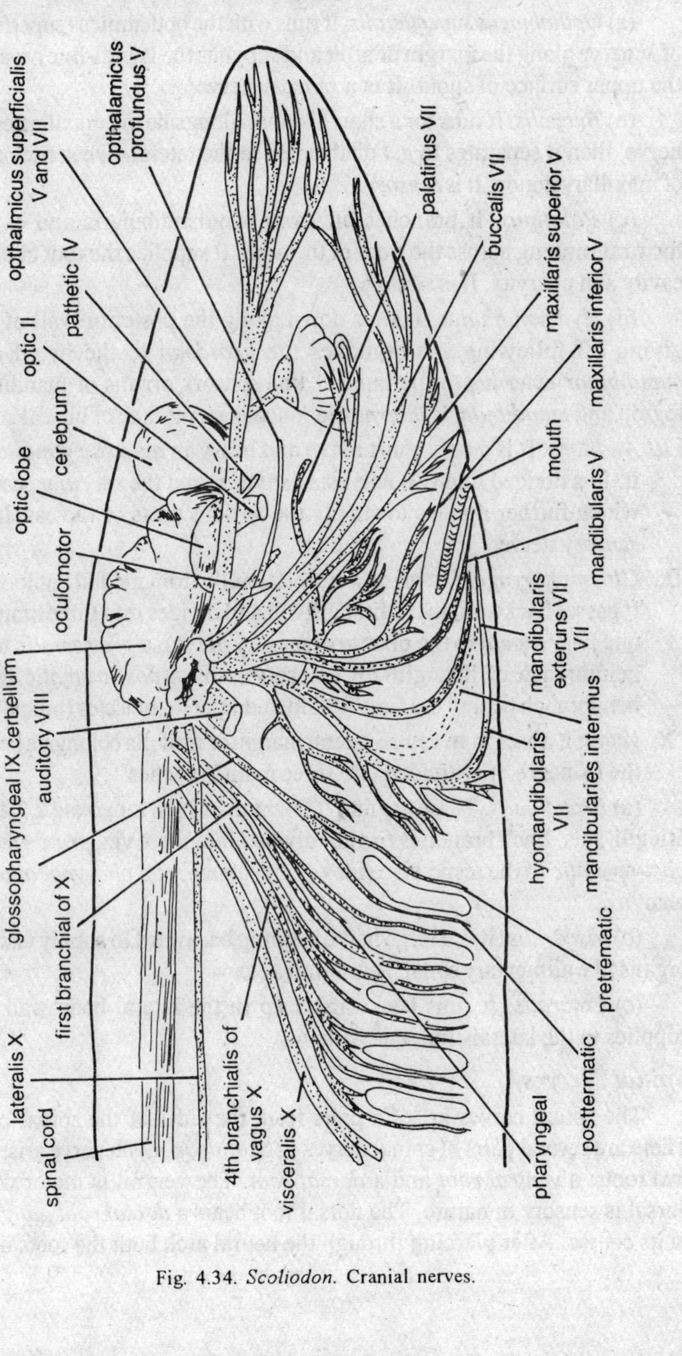

Fig. 4.34. *Scoliodon.* Cranial nerves.

(a) *Opthalmicus superficialis*. It runs with the opthalmicus superficialis of V nerve along the margin of orbit and supplies the lateral-line organs on the upper surface of snout. It is a *sensory* nerve.

(b) *Buccalis*. It runs for a short distance alongside the maxillaries of V nerve, then it separates to get distributed to the lateral-line sense organs of maxillary region. It is *sensory*.

(c) *Palatinus*. It branches off from hyomandibularis and extends forward running across the floor of the orbit. It supplies the roof of buccal cavity and pharynx. It is *sensory*.

(d) *Hyomandibular*. It runs down along the posterior wall of orbit giving off following sub-branches the *hyoidean* to the hyoid arch, *mandibular externus* to the latereal line sensory organs of mandibular region and *mandibularis internus* to mucous membrane of buccal cavity.

VIII. *Auditory*. It is a very short nerve and bears an *acoustic ganglion* on it. It is divided into two nerves; *vestibular* and the *saccular*, each of which further divides to supply the various parts of the ear. It is a *sensory* nerve.

IX. *Glossopharyngeal*. It arises ventro-laterelly from medulla oblongata. It passes backward to the first gill-slit and divides into three branches one *pharyngeal* to the pharynx (sensory nerve); a *pretrematic* to the hemibranch of first gill-slit (sensory); and a *post-trematic* to the hemibranch behind the first gill-slit and visceral muscles (mixed).

X. *Vagus*. It arises from ventro-lateral margin of medulla oblongata behind the IX nerve. It is divided into three main branches:

(a) *Branchialis*. There are four branchialis which innervate 2,3,4 and 5th gill-slits. Each branch is further divided into three viz. *pre-trematic*, *post-termatic* to the respective sides of gill-pouch, and *pharyngeal* to the pharynx.

(b) *Visceralis*. It is a large nerve running backward to supply visceral organs like alimentary canal, heart, lungs etc.

(c) *Lateralis*. It runs backward deep in the lateral body wall and supplies to the lateral-line sense organs.

Spinal Nerves

The spinal nerves arise in pairs from the sides of the spinal cord. There are several pairs of spinal nerves in *Scoliodon*. Each nerve arises by two roots: a *ventral root* and a *dorsal root*. The ventral is motor while dorsal is sensory in nature. The dorsal root bears a *dorsal root ganglion* in its course. After piercing through the neural arch both the roots unite outside to form the mixed spinal nerve. Each spinal nerve divides into three branches (i) *ramus dorsalis*, supplying the skin and muscles of the dorsal body wall, (ii) *ramus ventralis*, supplying the skin and muscles of the ventral body wall and (iii) *ramus communicans* containing the visceral sensory and motor fibres and joining the autonomic nervous system. A *brachial plexus* in formed at the level of pectoral girdle, but there is no plexus in the plevic region.

III. Autonomic Nervous System

It consists of a paired series of irregularly arranged ganglia situated anteriorly in the dorsal wall of the posterior cardinal sinuses and posteriorly in the dorsal part of the kidney on each side of the mid-dorsal line. The first ganglion is small but second is larger and remaining are small. Often there are two or more ganglia in each segment. The autonomic nervous system is not differentiated into sympathetic and parasympathetic systems in the fishes.

THE RECEPTOR ORGANS

The main receptor or sense organs are paired *olfactory*, *optic*, *ears*, *lateral line organs* and the *ampulla of Lorenzini*.

The Olfactory organs

There are a pair of blind sacs situated on the ventral side of the head infront of the mouth. Each olfactory sac is ellipsoid in shape and proportionately very large. It is covered externally by a thin membrane and is lodged on either side in the cartilaginous olfactory capsule. Internally they are lined with mucous membrane which is thrown into double series

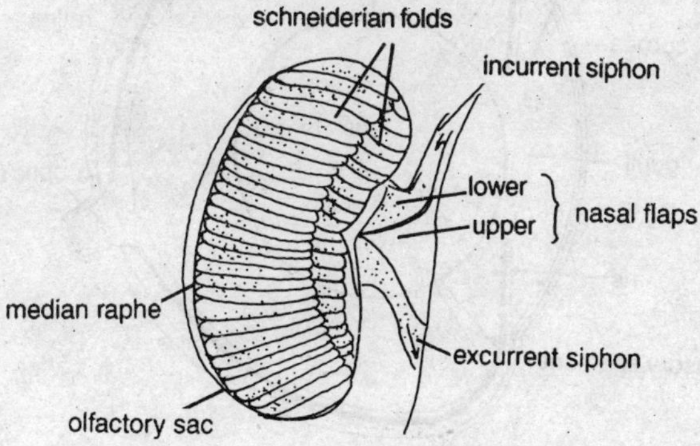

Fig. 4.35. *Scoliodon.* Olfactory organ.

of folds, the *Schneiderian folds*. These folds are held in position by a median connective tissue band called *median raphe*. The folds are covered by *olfactory epithelium*. The sensory cells of epithelium possess sensory hair at the free ends while their basal ends are continued into fibres of first cranial nerve. The olfactory sacs open out side by the external nares.

The Photoreceptors

There are a pair of large and well developed eyes. Each eye is provided with upper and lower eyelids and a nictitating membrane. These are elliptical in form and composed of three parts:

(i) *The fibrous tunic.* The fibrous tunic or sclerotic is cartilaginous and opaque, and forms the outer protective layer of the eye. Infront the sclerotic is transparent area. This area is known as *cornea*, through which light enters the eye ball. The cornea is covered by a transparent epithelium, the *conjunctiva*, which is continuous over the inner surface of the eyelids. Its part over,the cornea is called the *bulbar conjunctiva* and that lining eyelids is *palpebral conjunctiva.*

(ii) *The choroid.* The sclerotic is lined internally by a connective tissue coat, the *choroid* which is richly supplied with blood vessels and is pigmented. At the junction of the sclerotic and the cornea the choroid abruptly bends into the cavity of the eye in the form of circular partition,

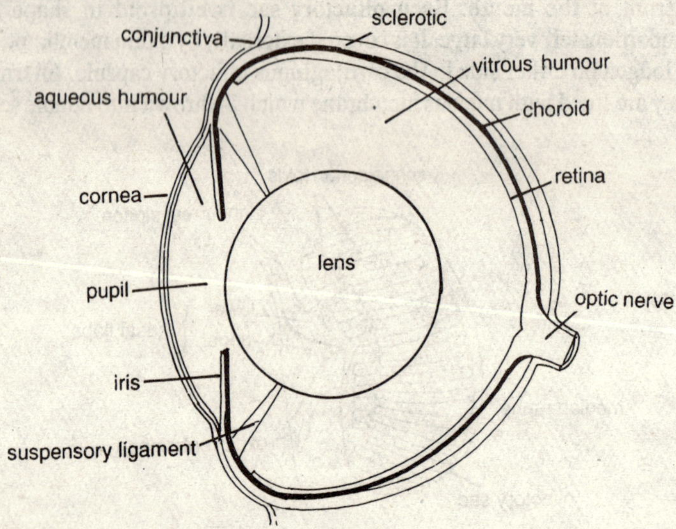

Fig. 4.36. *Scoliodon.* V.S. of eyeball.

the *iris*. The iris has a narrow vertical slit, called *pupil* in the centre. Behind the periphery of the iris, the choroid is thicker and less vascular and less pigmented known as *ciliary body*. In Elasmobranchii choroid contains cells with guanin crystals forming a light-reflecting surface called *tapetum lucidium*.

(iii) *The retina*. As usual, it has a sensitive and non-sensory *iridial* and *ciliary* parts. Only the rods are present, therefore, eye is adapted for dim-light vision. A *blind spot* and an *area centralis* are present but fovea centralis is absent.

A solid, spherical, transparent body, the *crystalline lens* lies just behind the pupil. It is held in the cavity of eye ball by *suspensory ligament* with *ciliary bodies*. A muscle, *protractor lentis* runs from the lower side of the lens to ciliary body, helps in accommodation for near vision.

The lens and suspensory ligament divide the cavity of the eye into two chambers the narrow anterior chamber is *aqueous chamber* which is filled with a clear watery fluid, the *aqueous humour*, and a spacious posterior chamber, the *vitreous chamber* filled with a gelatinous mass, the *vitreous humour*. The eye ball is provided with optic nerve which leaves the eye at blind spot.

The eye ball remains attached to the inner wall of orbit by six eye muscles and a certilaginous stalk called *optic pedicle*. There are two *oblique muscles* and four *recti muscles*.

Working of eyes. The eyes are laterally placed so that binocular vision is not possible, the vision is monocular. Accomodation is also poor because the pupil is not contractile as well as very little change occurs in lens due to protacter lentis muscle. In dim light the lens and humours refrect the rays of light passing through them to bring a focus on the retina. A shark is normally far-sighted, but as its retina is without cones. It is colourblind.

The Ears or Stato-acoustic organs

The ears are the organs of hearing and equilibrium. The external and middle ears are absent only the internal ears are present known as *membranous labyrinth*. Each internal ear is ectodermal, sac-like delicate organ enclosed in a cartilaginous auditory capsule. The main body or *vestibule* is laterally compressed and is made up of a dorsal *utriculus* from which projects a ventral and posterior lobe, the *sacculus*. The sacculus and utriculus are joined by the *sacculo-utricular duct*. The sacculus gives off a conical projection from its lower side known as *lagena-cochlea*. Besides the vestibule, there are three *semi-circular canals*. Each semicircular canal is dilated at one end into the *bulbous ampulla* but opens at both the ends into the vestibule. From the top of utriculus arise the *anterior vertical* and *horizontal semi-circular canals*, their ampullae lie close together and open anteriorly into the middle of the urticulus. Beneath these ampullae the utriculus gives off *recessus utriculi*. The sacculus dorsally gives origin to the *posterior vertical semi-circular canal* which forms a complete circle and opens posteriorly by its ampulla back into the lagena, which is the posterior out growth of the sacculus.

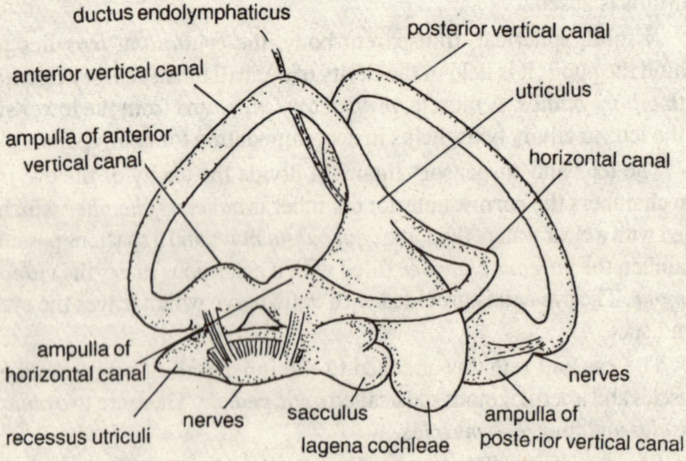

Fig. 4.37. *Scoliodon*. Membranous labyrinth.

The cavity of membranous labyrinth is filled with a fluid, the *endolymph* in which floats one or more calcareous bodies, the *otoliths*. The cavity of sacculus communicates with the exterior through a tube, the *endolymphatic duct*, which runs upward and pierces the roof of cranium within the parietal fossa. The space between the membranous labyrinth and the wall of auditory capsule is filled with *perilymph*, the perilymphatic space opens to the exterior by a large aperture, the *fenestra*. The membraneous labyrinth is innervated by auditory nerve.

The membranous labyrinth performs two important functions (i) *Static function*. The semi-circular canals and utriculus are responsible for equilibrium i.e. maintain the balance of body. (ii) *Acoustic function* is performed by the sacculus and its lagena to receive the auditory stimuli.

LATERAL LINE SYSTEM OR NEUROMAST SYSTEM

The lateral line system is present in fishes, cyclostomes, some aquatic amphibians and tadpole larva of amphibians. This system includes (A) Lateral lines receptors, (B) Neuromast organs and (C) Pit organs.

A. Lateral-line receptors

The lateral-line receptors are situated in the lateral-line canal on each side of the trunk and tail. Each lateral-line canal is continued into the *cephalic canal* on the head. These canals are situated in the dermis and possess groups of epidermal receptors or *neuromasts*. On reaching the head each lateral-line canal bends dorsally and is joined with the canal of opposite side by a transverse commissure, the *transverse commissural occipital canal*. It is situated immediately behind the opening of endolymphatic duct. The commissural occipital canal is divided into two canals:

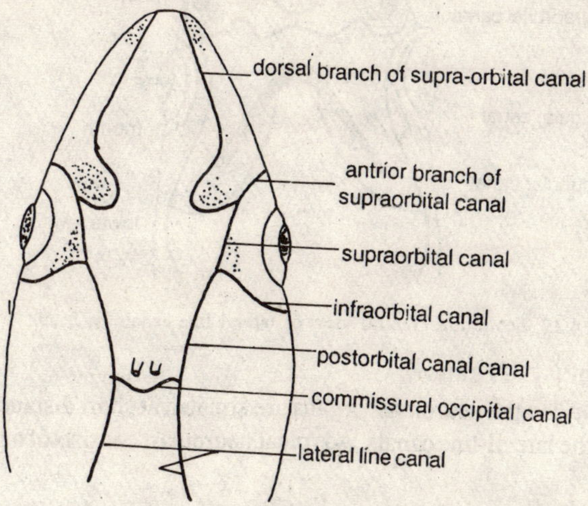

Fig. 4.38. *Scoliodon*. Dorsal view of lateral line canals on head.

(i) *Supra orbital* which runs forward, dorsal to the eye and is divided into two branches a *dorsal branch* and an *anterior branch*. The dorsal branch runs to the tip of the snout and bends sharply backward as the ventro-lateral branch running along the ventro-lateral border of snout. The anterior branch bends downwards just infront of the eye and joins the infra-orbital.

(ii) *The infra-orbital* runs along the posterior margin of the orbit. It curves down to the ventral surface and joins the anterior branch of supra orbital canal. Later on, it curves ventrally and joins the ventro-lateral branch of *superior-orbital* and runs transversely inwards to meet its fellow of the opposite side infront of the palatopterygo-quadrate symphysis. The fused-infra orbital canal runs forward a short distance, in the mid-ventral line, but separates again and extends parallel to the other upto the tip of snout to join the supraorbital canal. Another canal called *jugal canal* arises from the place of union of the infra-orbital canal and ventro-lateral branch of supra-orbital canal. The jugal canal gives off a mandibular canal at the angle of the jaws. This canal runs upto the anterior end of the lower jaws.

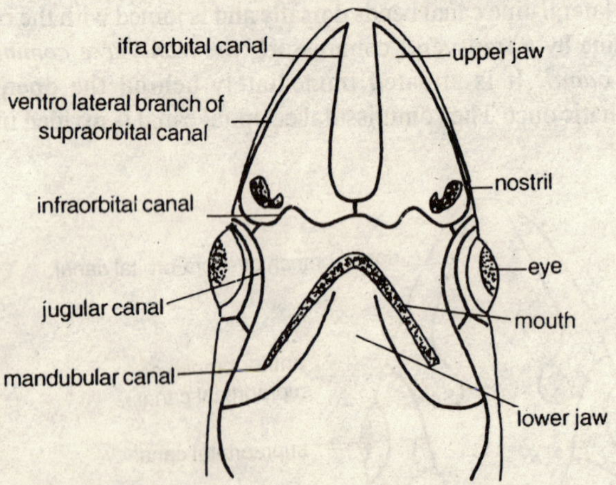

Fig. 4.39. *Scoliodon.* Ventral view of lateral line canals on head.

B. Neuromast Organs

In *Scoliodon* the neuromast organs are situated at short distances on the floor of the lateral-line canals. A typical neuromast consists of a group

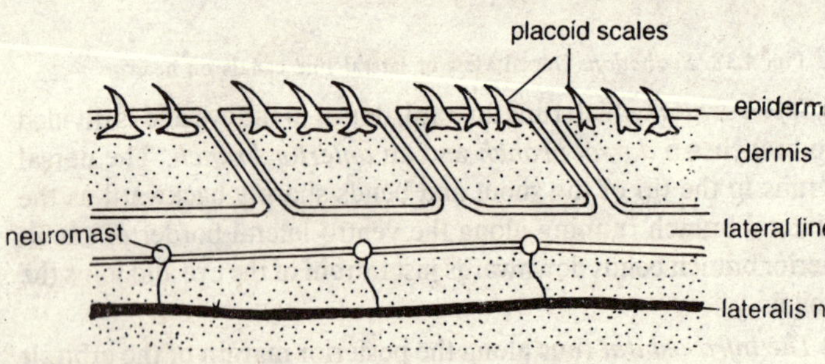

Fig. 4.40. *Scoliodon.* Lateral-line canal in longitudinal section to show the arrangement of neuromast organs.

of *neuro-sensory cells*, each with a sensory hair at one end and a nerve fibre at other end. The neurosensory cells are supported by *supporting cells*. The sensory cells are innervated by the nerve fibres of VII, IX and X cranial nerves. These sensory cells are stimulated by the movements occuring in the fluid found in the lateral-line canals. They can preceive vibration of low frequency and detect disturbance in water caused by the movements of other fishes. Thus these are *rheoreceptors*.

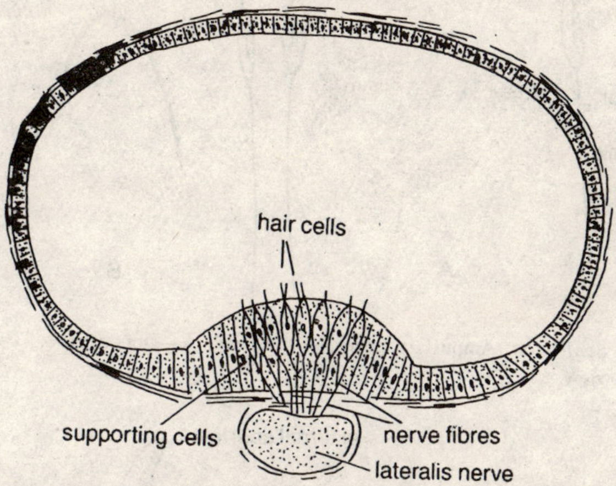

Fig. 4.41. *Scoliodon.* T.S. of lateral line canal.

C. Pit-organs

These are also called *prototriches*. They are present on the dorsal and lateral surfaces of head. They are embedded in pits. A pit-organ consists of a few neurosensory cells and supporting cells sunk into a small pit in the skin. They are also *rheoreceptors* as they respond to vibrations in the water.

Ampulla of Lorenzini

These are definite groups of minute pores on the dorsal surface of the head of *Scoliodon*. Each of these pores leads into an elongated tube which ends in a radially separate ampullary sac lying deeply beneath the skin and free from it.

According to their location, the ampullae are of three types; the *supra-opthalmic group* lies around the supra-orbital canal; the *outer buccal group* between supra-orbital and infra-orbital canals; and the *inner buccal group* beneath the infra-orbital canal.

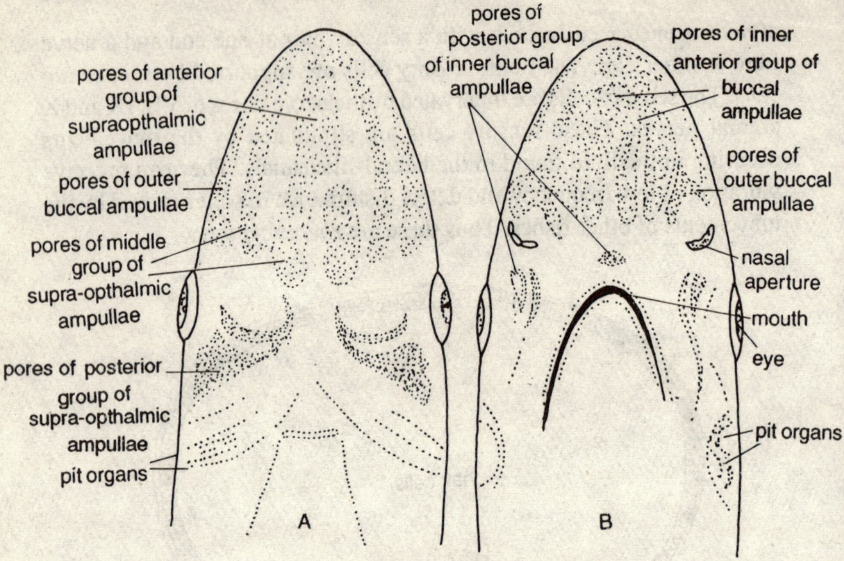

Fig. 4.42. *Scoliodon.* Ampullary pores on the head. A—Dorsal view. B—Ventral view.

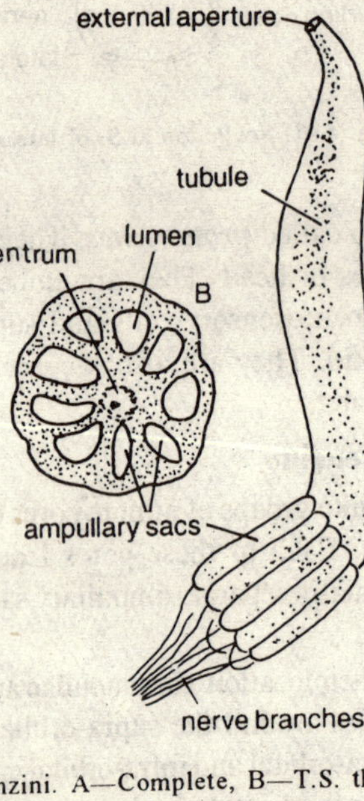

Fig. 4.43. Ampulla of Lorenzini. A—Complete, B—T.S. through the ampullary sac.

Each ampulla consists of eight or nine radially dilated chambers arranged around the central core known as *centrum*. There are two types of cells present in ampullae : the *pear-shaped gland cells* and *pyramidal cells.* The pear-shaped glands secrete the jelly filling the ampullary tube. The pyramidal cells bear sensory hairs forming sensory cells. Each ampulla is supplied by nerve fibres from VII cranial nerve.

Ampulla of Lorenzini were regarded aș neuromast organs. But the studies by *Sand* (1938) shows they are the *thermoreceptors* responding to change in temperature of surroundings.

URINOGENITAL SYSTEM

Like other vertebrates, the excretory and reproductive systems are closely related and associated to each other so that they are studied under the same heading 'urinogenital system'. In *Scoliodon*, the sexes are separate and the urinogenital system includes different organs in the male and female fishes.

Male Urinogenital System

The excretory part of system consists of a pair of kidneys, a pair of urinary ducts and a urinogenital sinus.

Kidneys

The kidneys are *mesonephric* in origin. They are long, ribbon-like glandular structures lying dorsally to the peritoneum. They extend from the root of the liver in front to the side of cloaca behind. Each kidney is made up of two parts: an anterior narrow genital part or *Leidig's gland* and a posterior *renal part*. The substance of the kidney is made up of coiled urinary tubules, each of which consists of a peritoneal funnel, a *Bowman's capsule* enclosing the *glomerulus* and a coiled *renal tubule*. Several renal tubules open into a *common collecting tubule*. The collecting tubules of the anterior part of kidney are very small and they open into the *Wolffian duct*.

Ureters or urinary ducts

The collecting tubules of the posterior part of each kidney opens into a common duct, the *ureter* but it is not homologous with the unreter of amniotes where these are metanephric. The ureter runs backward and opens behind into the urinogenital sinus.

Urinogenital sinus

It is a large, somewhat triangular median sac lies at the end of body-cavity. It communicates with cloaca through an aperture, situated at the tip of *urinogenital papilla.*

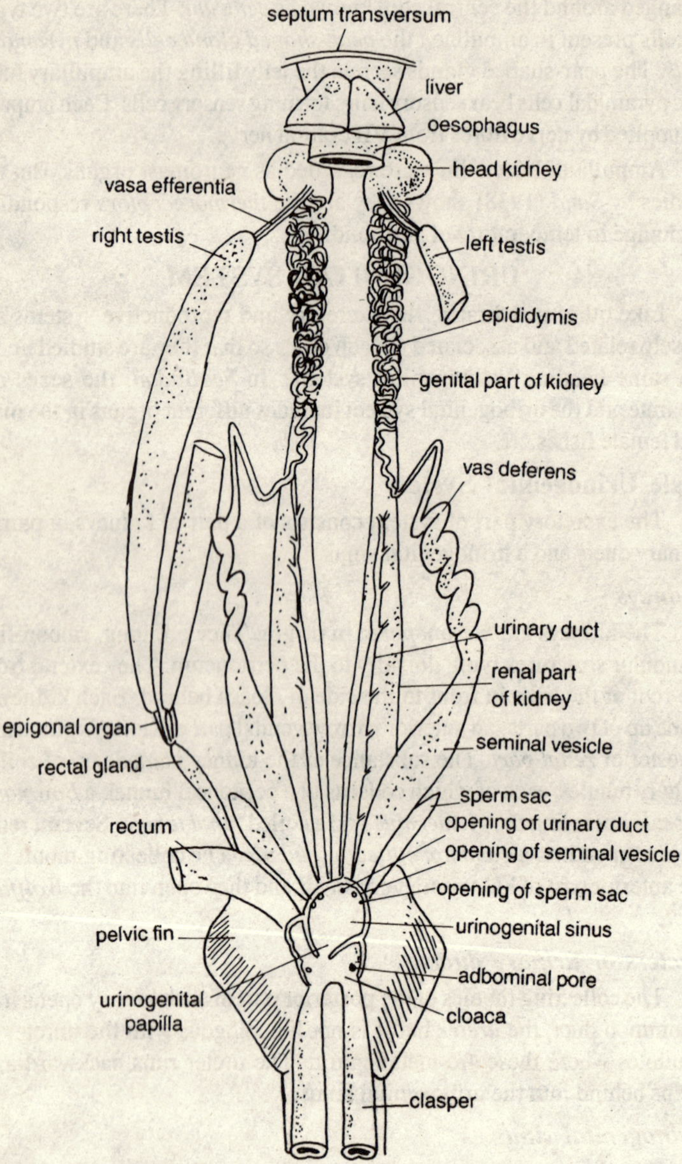

Fig. 4.44. *Scoliodon.* Male urinogenital system.

The genital part consists of testes, vasa efferentia, vasa deferentia or Wolffian ducts, sperm sac, siphon and claspers.

Testes

The testes are a pair of elongated structures made up of numerous lobules, the *seminiferous tubules* with a median central canal. They extend anteriorly from the base of liver to the rectal gland posteriorly. They are attached to the dorsal body wall by folds of peritoneum called the *mesorchia*. Posteriorly, they are connected with rectal glands by means of non-glandular *epigonal organs*. The seminiferous tubules are lined with germinal epithelium.

Vasa efferentia

The vasa efferentia are fine tubules arising from the central cavity of each testis. They run in the mesorchium to the anterior end of a large Wolffian or mesonephric duct. The Wolffian duct now is called vas deferens.

Vasa deferentia

Each vas deferens passes along the ventral surface of the genital part of the kidney. It is highly convoluted in the adult, forming the *epididymis*. Reaching the renal part of kidney, the Wolffian duct becomes much wider and much less convulated called *vasicula seminalis* for the storage of sperms. The vasicula seminalis of two sides open behind into a common urinogenital sinus.

Sperm Sacs

On either side of urinogenital sinus lies a club-shaped sperm sac which is really an outgrowth of urinogenital sinus. The function of these sacs is unknown.

Siphons

The siphons are the accessory part of reproductive system. These are a pair of elongated and glandular sac-like siphons, present on the ventral side under the skin which end blindly at the base of pelvic fin. They communicate behind with the grooves of claspers. They are full of sea-water and serve to force to sperms into the female genital tract by way of the claspers.

Claspers

The claspers are stiff, rod-like copulatory organs and are supported by cartilages. Each clasper has a dorsal groove, whose edges are folded scroll-like, converting the groove practically into a tube. The tube is open at both the ends. The anterior opening is known as *apopyle*, communicate with cloaca as well as siphon and receives sperms. The posterior opening is called *hypopyle* which serves as the exit through which sperms leave the clasper and pour into the cloaca of female.

Femal Urinogenital System

The female urinogenital system differ from those of the male in that there is no direct connection between the kidneys and the genital parts and that the anterior part of each kidney is extremely reduced. The excretory part of the system includes:

Kidneys

The kidneys are mesonephric and they occupy the same position as in the males. The anterior part is reduced to narrow tapering strand. The excretory tubules form the posterior part of kidney open into the mesonephric or Wolffian duct, called ureter.

Ureters

The mesonephric duct from each kidney runs backward on the ventral surface and joins its fellow to form a single median duct, which opens into the urinary sinus.

Urinary sinus

It is a large median sac lying at the end of body cavity. It communicates behind with the cloaca by an aperture situated on a urinary papilla.

The genital part of the system includes:

Ovaries

A pair of ovaries are situated one on either side of the vertebral column, just behind the base of the liver. They are suspended in body cavity by a fold of peritoneum, the *mesovarium*. The ovaries varying in shape and size according to age of fish. Both the ovaries are conncted with the anterior end of the rectal gland by *epigonal organs.*

Oviducts or Mullerian ducts

The oviducts begin from the septum transversum and extend throughout the entire length of the body cavity uniting behind to form *vagina*. The vagina opens into urinogenital sinus. The anterior end of oviducts reach the mid-dorsal line near the base of the liver, so that the two oviducts communicate with the coelom by a single large median longitudinal slit, the *ostium*. Each oviduct then runs backward and enlarges a short distance to form the *shell gland*. Finally the oviduct dilates to form a sac-like structure, the *uterus*. Two uteri unite to form *vagina*.

The oviducts and uteri are held in the body cavity by paritoneal fold, the *mesotuberium*.

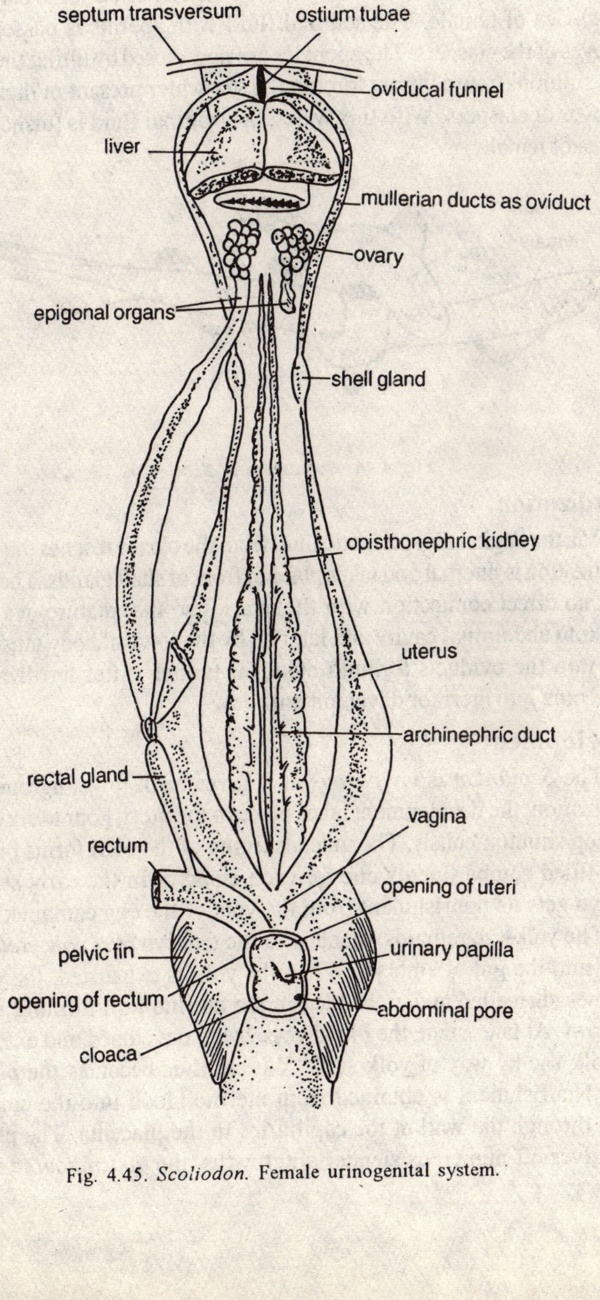

Fig. 4.45. *Scoliodon*. Female urinogenital system.

COPULATION

The reproduction occurs throughout the year. During copulation the male twines round the female. It inserts its one or both the claspers into the cloaca of female. The seminal fluid with sperms is passed into the grooves of the claspers. The apopyles are now closed by tilting the claspers. The siphions now contract forcing the sea-water present in them into the grooves of claspers. With this water, the seminal fluid is flushed into the cloaca of female.

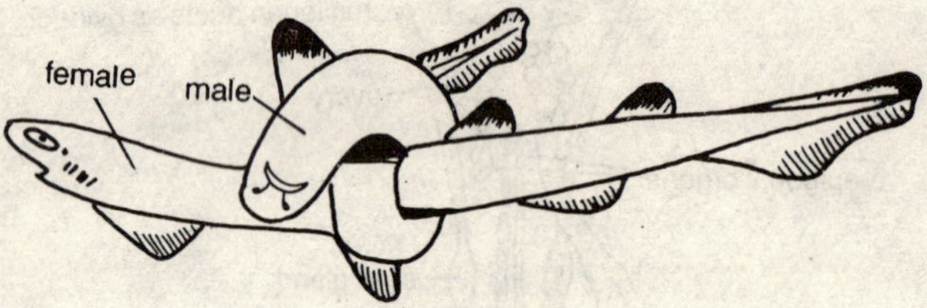

Fig. 4.46. Copulation in *Scoliodon*.

Fertilization

Via the vagina the spermatic fluid from the cloaca reaches the oviducts. Fertilization is internal and takes place in front of shall glands. The oviducts have no direct connection with the ovaries so that mature ova are first shed into abdominal cavity and later on by the force of body muscles they pass into the oviducts through oviducal funnel. After fertilization the eggs come into uteri for development.

Development

The *Scoliodon* is *viviparous* as it gives birth to young ones. After fertilization, the development takes place in the uteri. Four to six embryos develop simultaneously. The mucous lining of the uteri forms protective fluid-filled compartments one for each embryo. In the early stages the embryo gets its nourishment from the yolk of the egg contained in *yolk sac*. The yolk-sac remains attached to the embryo by a *yolk stalk* which leads into the gut of embryo. When the yolk is exhausted, the yolk sac becomes shrivelled and embedded in the uterine wall forming *yolk sac placenta.* At later stage the blood vessels are developed and extend into the yolk sac by way of yolk stalk. Which, thus becomes the *placental cord.* Nourishment is obtained from uterine blood into the embryonic blood through the wall of the capillaries in the placenta. The placental cord gives off numerous slender outgrowths, the *appendiculae* into the

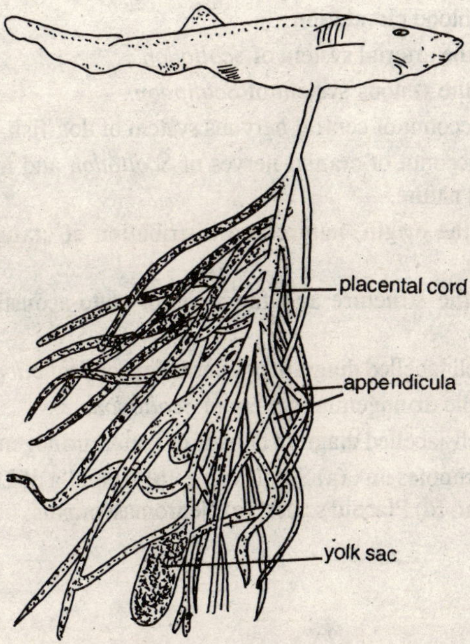

Fig. 4.47. *Scoliodon*. Embryo with placenta.

uterus to absorb nourishment from uterine wall. The young ones born are about 15 cm in length.

Revision Questions

1. Give an account of structure and development of placoid scales.
2. Describe the habits, habitat and external morphology of *Scoliodon.*
3. Give an account of digestive system of *Scoliodon* and explain the process of digestion.
4. Give an account of respiratory system in *Scoliodon.* Give mechanism of respiration.

5. What is venous heart ? Describe the internal structure of heart and course of blood circulation.
6. Describe the arterial system of *Scoliodon.*
7. Describe the venous system of *Scoliodon.*
8. Give an account of central nervous system of dog-fish.
9. Give an account of cranial nerves of *Scoliodon* and mention their origin and nature.
10. Describe the origin, nature and distribution of cranial nerves in *Scoliodon.*
11. Describe the structure and functions of stato-acoustic organs of *Scoliodon.*
12. Draw a well labelled diagram of internal ear of *Scoliodon.*
13. Describe the urinogenital system of *Scoliodon.*
14. Draw a well-labelled diagram of male or female urinogenital system.
15. Write short notes on : (a) Scroll valve, (b) Ampulla of Lorenzini, (c) Internal ear, (d) Placoid scales, (e) Neuromast organs.

5

Labeo rohita (Rohu, A Bony Fish)

There are about 20,000 living species of bony fishes. They vary greatly in shape and proportions, but they are built on the same basic plan. Carps and cat-fishes are common fishes in the quite, clear and sluggish rivers and estuaries of Indian sub continent. *Labeo rohita* is the most famous carp, and serves well as a type of the bony fishes. Its local name is *Rohu*, but is also called '*Dambra*' or '*Tapra*'. The fishes may occur in ponds also. About two dozens of species are known from India. Among these *L. rohita* (Rohu), *L. calbasu* (Kalbasu), *L.kontius, L. gonius, L. nandina* and *L. fimbriatus* are common. Out of these, Rohu and Kalbasu have been more relished as food fishes.

Habits and Habitat

Labeo is carnivorus when young, and feeds upon *Daphnia* or water fleas and insect larvae. It becomes herbivorous when adult and eats up algae and aquatic plants. Occasionally even the adult *Labeo* can consume the animal diet. Because of the feeding habits, it is usually cultivated in association with two other carps, *Catla catla* (Catla) and *Cirrhina mrigala* (Mrigal), the two other important food fishes, by the fish culturists. It prefers clean water and for respiration it develops gills. Sexes are separate but there is no sexual dimorphism. The fish is oviparous, and breeds only in the rivers but not in a confined waters. Fertilization is external. Its breeding period is from July to August. *Labeo* frequently comes to water surface to take air into the air bladder.

EXTERNAL FEATURES

The body of *Labeo* is elongated, laterally compressed, spindle shaped tapering at either end. The colour of the body is dark greyish on the back, but sides and belly are pale yellow or white in colour. The fish grows about a metre and weights about ten kilograms. The body is divisible into head, trunk and tail.

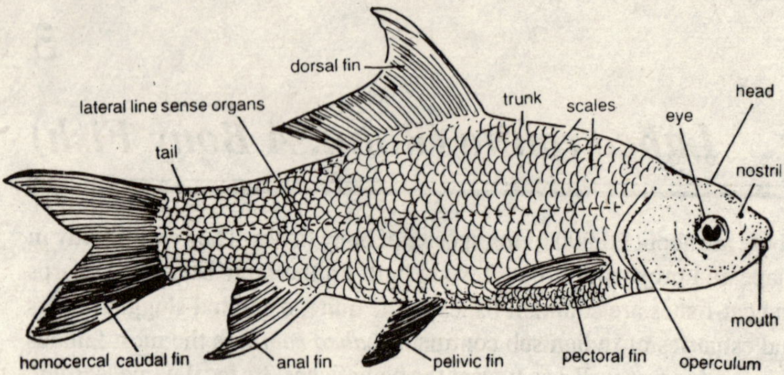

Fig. 5.1. *Labeo*. External Features.

Head

The head is depressed and is produced anteriorly as short, obtuse, depressed and swollen snout. This makes the mouth sub-terminal instead of terminal. Mouth is large transverse aperture bounded by thick and fleshy lips. There is a pair of short, thread like, tapering, sensory processes on the snout, one on either side just above the mouth. These are the *maxillary barbels*. The jaws are without teeth, instead, there are three rows of teeth in the pharynx. The paired eyes are large, without eyelids but a transparent membrane covers the cornea. The pupil is rounded. There are two nostrils situated dorsally in front of eyes. They open internally into the olfactory sacs. There is a large, bony flap below and behind each eye. This is the *operculum* which is also called the *gill cover*. Between the gill-slits and the operculum lies a spaceous *branchial cavity* or *opercular chamber*, which communicates with the exterior by a single aperture, which is crescentic and can be shortened and widened by the *branchiostegal fold* that borders the opening. There is no operculum in sharks, The mechanism of breathing and gill structure of rohu is quite different from that of the sharks.

Trunk

It extends from the posterior borders of the operculum to the anal aperture. It is thickest part of the body and is compressed laterally. On either side of trunk and tail is a *lateral line*. The trunk is covered by the dermal *cycloid scales* which are rounded with free edges and not spiny.

They are arranged lengthwise and differ from those of sharks in being true bony, not provided with enamel. The trunk bears five fins including paired and unpaired fins. The fins are supported by bony *fin rays*. The single *dorsal fin* is some what rhomboidal and is supported by 12-13 jointed bony fin rays. The *anal* or *ventral fins,* lying just posterior to the anus and each has eight fin rays. Amongest the paired fins, the pectoral fins are found ventro-laterally behind the operculum. The pectoral fin bears 19 fin rays each. The pelvic fins are found more or less in the middle of the body, lying ventro-laterally, and have nine fin rays each. There are three apertures in the trunk region. These include the *anal, urinary* and *genital apertures.* The anal aperture is the posterior opening of alimentary canal and lies on the ventral side just infront of the anal fin. The urinary aperture is present behind the anal aperture. The genital aperture lies between the anal and urinary openings.

Tail

The tail forms about one third of the body. It is laterally compressed and narrower behind. The *caudal fin* arises from the hind end of the tail and is deeply forked into two similar lobes. The *epicaudal* and *hypocaudal lobes* are roughly equal. Thus the tail fin is *homocercal.* There are nearly nineteen fin rays in the tail fin of rohu. The movement of this fin pushes the fish to move forward.. The tail makes the principle locomotory organs. Tail is also covered over by dermal, cycloid scales like the trunk. The cloaca is absent. Copulatory organ are also not found.

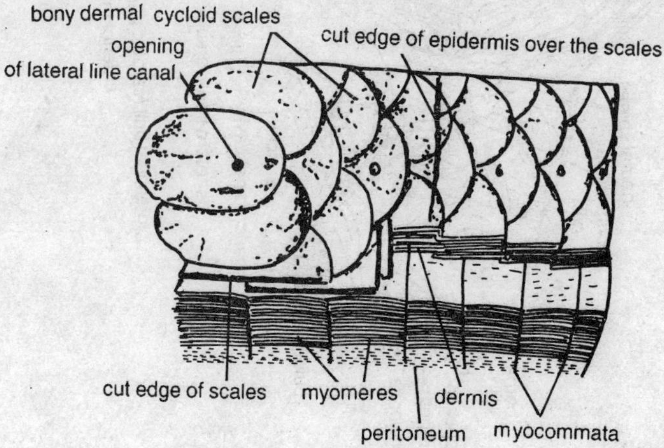

Fig. 5.2. *Labeo.* Body wall.

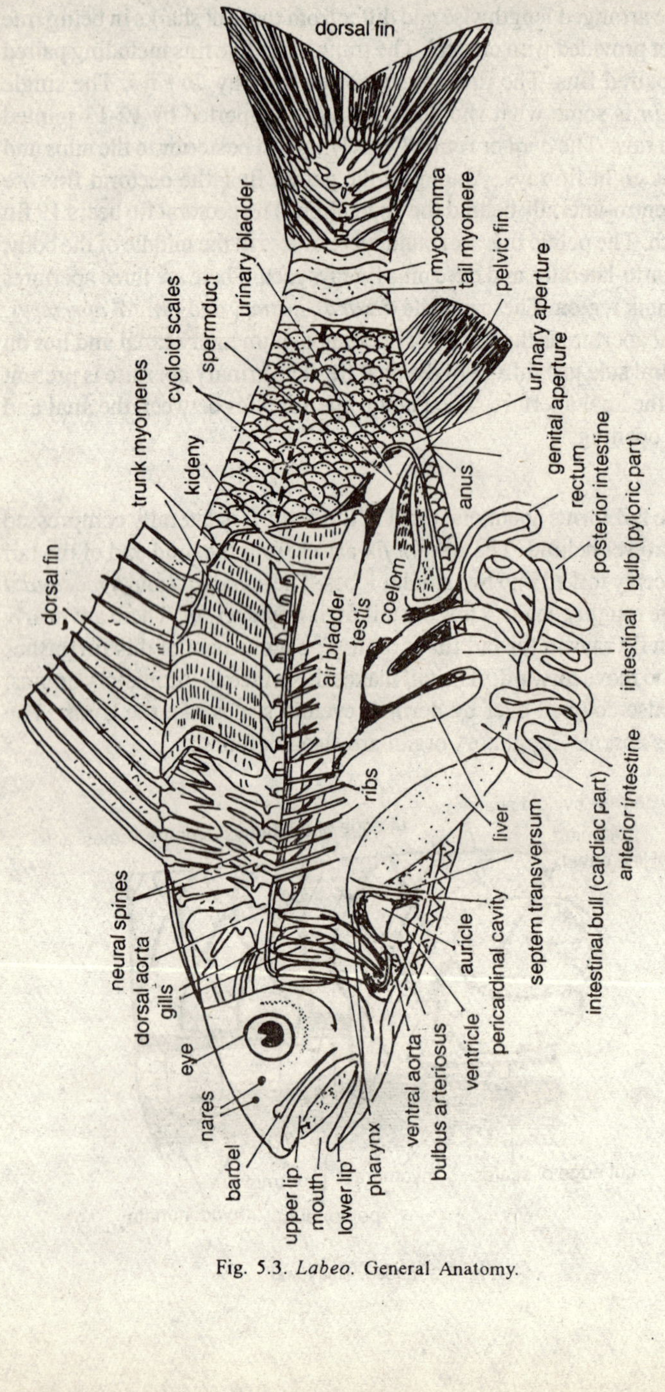

Fig. 5.3. *Labeo*. General Anatomy.

BODY WALL

The skin of trunk and tail contains rows of dermal scales. These are the cycloid scales which are set at an angle in the skin so that they overlap and form a complete covering. A part of each scale is burried in the dermis and points forward. Overlapping arrangement allows the scale to slide over one another, and provides more flexibility.

The skin is differentiated into two parts; upper thin *epidermis*, and lower thick and muscular *dermis*. The epidermis is made up of stratified epithelium and rests on a *basement membrane*. There is no horny layer of flat, dead cells characteristics of land vertebrates. The epidermis contains *mucous glands* that secrete mucous to make the skin slimy and frictionless. The dermis is mainly composed of connective tissues. The fibres remain parallel to the surface. Pigmented cells are present between the epidermis and dermis. Cycloid scales are lodged in the pockets of the dermis. The principle muscles of trunk and tail are arranged is zigzig myotomes as in other fishes and vertebrates.

ENDOSKELETON

In *Labeo* the endo-skeleton is mainly bony and is well differentiated into *axial* and *appendicular bones*. The axial part includes *skull, vertebral column* and *ribs*, whereas the appendicular part includes *pelvic* and *pectoral girdles* with their related *fin skeleton*.

Skull

Skull is cartilagenous in the beginning and then after undergoes ossification and turns bony. The skull comprises *cranium* and *visceral skeleton*.

The posterior most part of cranium is the occipital region which is made up of four bones : *supra occipital, basioccipital* and two *exoccipitals*. In addition there are three apertures in the occipital region a small median *foramen magnum* and two large oval *fenestrae* characteristic of cyprinoid skulls. Dorsally supraoccipital is overlapped by the parietals which lie in front. The supraoccipital bears dorsally a median longitudinal vertical occipital spine. Basioccipital is found below the two lateral exoccipitals. The dorsal process of exoccipital meets that of the other side to enclose the foramen magnum dorsally.

The otic region includes four bones namely the *prootic*, the *epiotic*, the *sphenotic* and the *pterotic*. There is no opisthotic in adult *Labeo*. A small bean shaped *supratemporal* bone is present between pterotic and the posterior process.

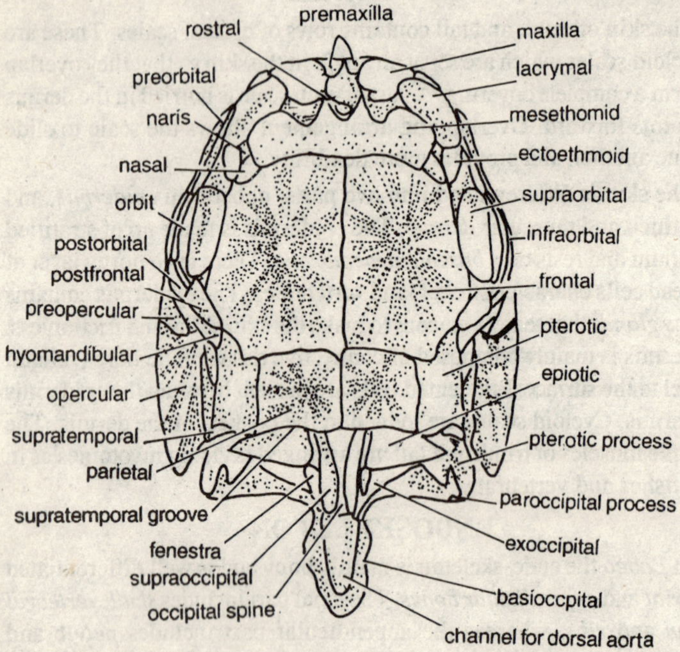

Fig. 5.4. *Labeo*. Skull (Dorsal view).

The major part of roof of skull is composed of *parietals* and *frontals*. There is a pair of parietals above, *paraspheonid* below and paired *alisphenoids* on the side. A pair of *frontals* above, *parasphenoid* below and paired *orbitosphenoids* on the sides.

The floor of the skull is a composed of parasphenoid. There is no basisphenoid. The frontal articulates with *mesethmoid* medially, the *nasal* and the *ectoethmoid* laterally, *supraorbital*, anterio-laterally and *postfrontal* postero-laterally.

The orbits are found in the frontal region and are bounded dorsally by frontal ventrally by alisphenoid and orbitosphenoid, anteriorly by ecto-ethmoid, and posteriorly by sphenotic.

The nasal or ethmoidal region is composed of nine bones. This region includes paired *nasals* above the olfactory capsule, median *mesethmoid* infront of frontal, a median *rostral* to the anterior concave notches of the mesethmoid and a ventral median *vomar* lying infront of parasphenoid and beneath the mesethmoid. *Ectoethmoid* and *lacrymal* bones also belong to this region.

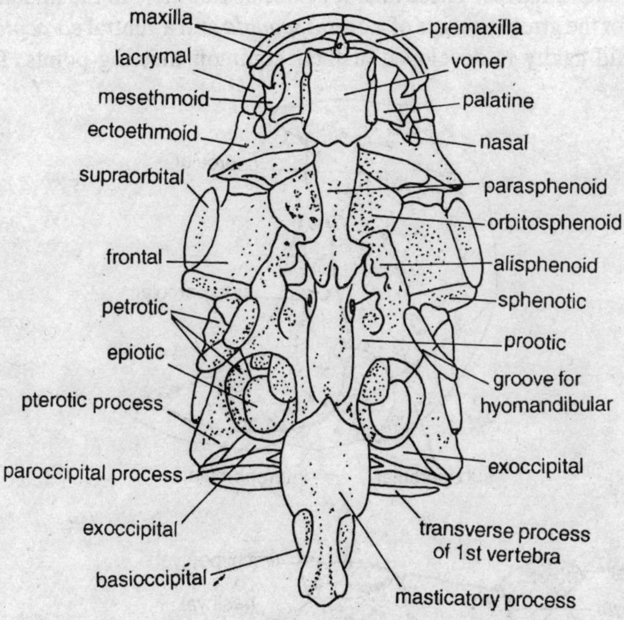

Fig. 5.5. *Labeo*. Skull (Ventral view).

Visceral Skeleton

The visceral skeleton is made up of several pairs of visceral arches. Among these, first pair is the *mandibular* second pair is the *hyoid* and five *branchial arches*. The mandibular arch form the upper and lower jaws. The upper jaw consists of *premaxilla, maxilla, palatine, metapterygoid* and *quadrate*. Similarly lower jaw consists of *dentary, angular* and *articular* bones. Teeth are never developd on these jaws bones in *Labeo*. The hyoid arch is formed of hyomandibular dorsally and the epihyal, the ceratohyal and the hypohyal ventrally. The posterior corner supports the tongue and the floor of mouth cavity. The last five pairs of visceral arches are called the gill or branchial arches. First four branchial arches bears gill lamellae while the last one is without gill lamellae. Typically, each arch is made up of four components; hypobranchial (ventral), ceratobranchial (ventro-lateral), epibranchial (dorso-lateral) and pharyngobranchial (dorsal).

Appendicular Skeleton

It includes mainly the bones of *pectoral* and *pelvic girdles*.

The *pectoral girdle* lies just behind the last branchial arch. It is made up of two lateral halves. These lateral halves do not meet in the middle. Each half of the girdle consists of a dorsal *scapula* and a ventral *coracoid*. The glenoid cavity is developed at their common meeting points. In

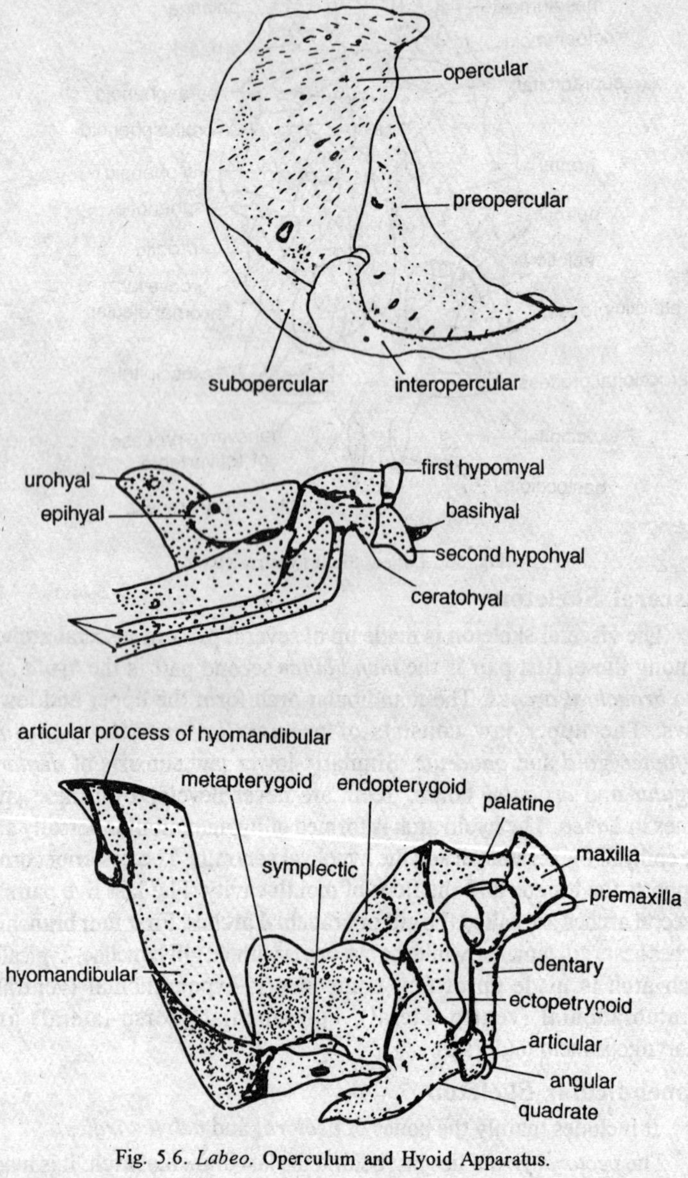

Fig. 5.6. *Labeo.* Operculum and Hyoid Apparatus.

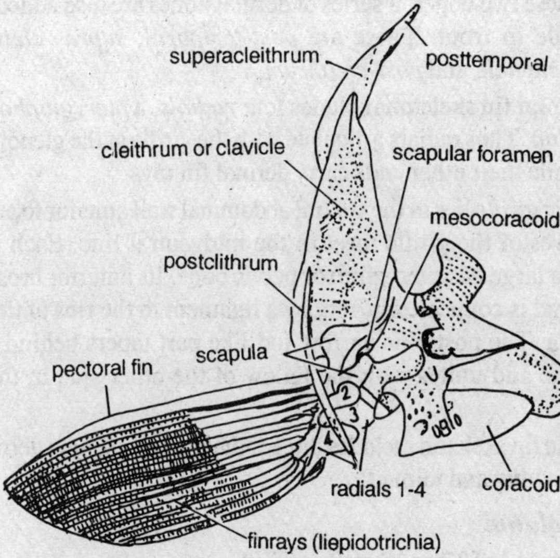

Fig. 5.7. *Labeo*. Pectoral fins and girdle.

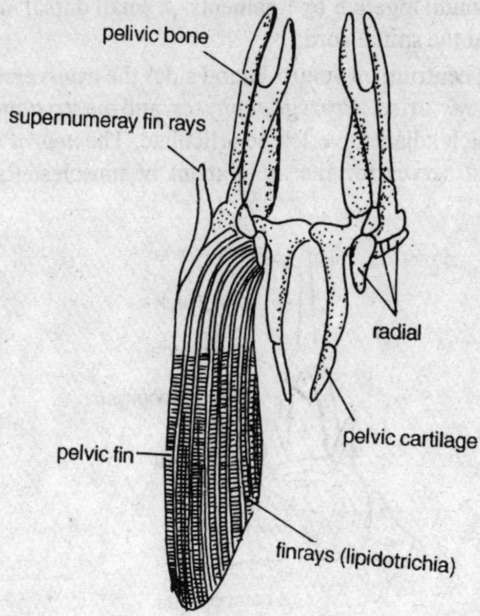

Fig. 5.8. *Labeo*. Pelvic fins and girdle.

addition to these two bones, a series of dermal bones are also added to the pectoral girdle in front. These are *post-temporal*, *supra- cleithrum*, *cleithrum* or *clavicle* and *post-cleithrum*.

The pectoral fin skeleton includes four *radials*, a *pterygiophore* and 19 *lepidotrichia*. Thus radials articulate with the girdle at the glenoid facet at one end while their other ends carry dermal fin rays.

The *pelvic girdle* lies in the ventral abdominal wall anterior to anal fin. Both the halves of the girdle meet in the midventral line. Each half is composed of a large basipterygium or pelvic bone. Its anterior broad part with forked end is connected infront by a ligament to the ribs of the 12th trunk vertebra. The posterior narrow rod like part tapers behind into a small cartilage and unites with the fellow of the other side in the mid ventral line.

The pelvic fin skeleton includes three radials, carrying nine dermal fin rays which are long and jointed.

Vertebral Column

It is composed of 37 or 38 well ossified *amphicoelous* vertebrae.

1. *Trunk vertebrae*. From 5 to 17 or 18 are alike called *typical vertebrae*.The centrum is short, bony and concave at both the ends. The adjacent are bound together by ligaments. A small dorsal *neural arch* is present around the spinal cord.

From the centrum on ventro-lateral sides the *transverse proccesses* or *parapophyses* arise. *Prezygapophyses* and *postzygapophyses* are present by which adjacent vertebrae articulate. The *neural spine* is well developed and serves for the attachment of muscles. Ribs are well

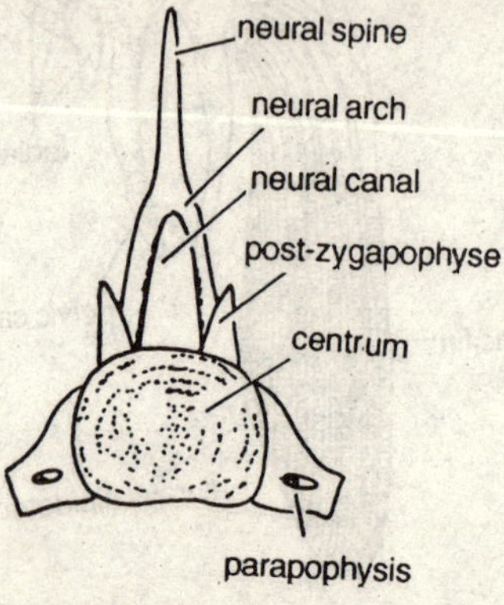

Fig. 5.9. *Labeo*. Trunk vertebra.

developed and remain attached to the tranverse processes of trunk vertebrae.

2. *Caudal vertebrae*. A typical caudal vertebra basically has the same parts as that present in trunk vertebra, except that there are no ribs, and on the ventral side there is a *haemal arch* which is produced into the *haemal spine*. The caudal vein and artery pass through the haemal canal.

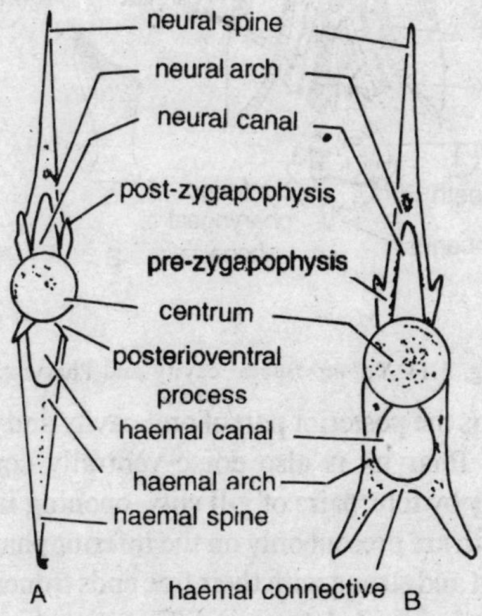

Fig. 5.10. *Labeo*. (a) Typical caudal vertebra (posterior view) and (b) First caudal vertebra (anterior view).

DIGESTIVE SYSTEM

Labeo rohita is *herbivorous* as far as its feeding nature is concerned. The digestive system includes alimentary canal and digestive glands :

Alimentary canal. It is complete and tubular and is divisible into four zones, in the usual way :

I. Ingressive zone

The part of gut ment basically to receive food. It includes mouth, buccal cavity, and pharynx.

The *mouth* is crecentic and subterminal and is bounded by felshy lips. The lips bear pigmented sensory papillae containing taste buds. The fish has no teeth on the edge of the jaw. Teeth are placed on the pharyngeal floor. The mouth leads into the buccal cavity.

Buccal cavity is the anterior part of oral cavity. It is quite spacious. It is compressed dorso-ventrally but wide laterally. Distinct tongue is absent.

Its inner lining which is made up of mucous membrane is raised into minute papillae. The buccal cavity leads into pharynx.

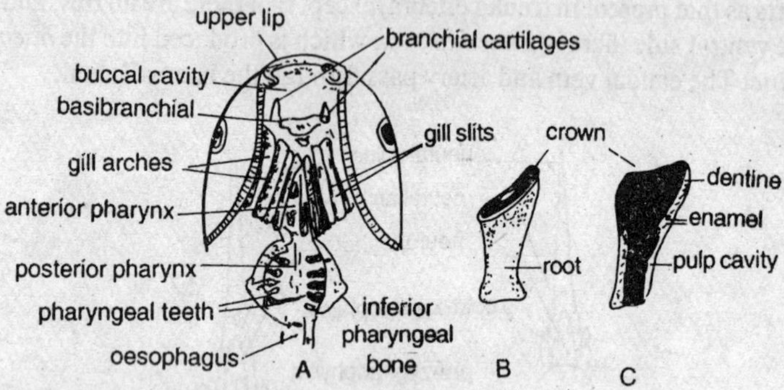

Fig. 5.11. *Labeo.* Buccal cavity and Pharynx.

The *pharynx* is the posterior part of oral cavity and is present next to the buccal cavity. Pharynx is also dorso-ventrally compressed and is perforated laterally by four pairs of gill slits, opening into the branchial chambers. The teeth are present only on the inferior pharygeal bone. The teeth are homodont and strong with their free ends truncate. The teeth are *acrodont* and *polyphyodont*. A tooth is differentiated into two parts : the *root* and the *crown*. The root is embedded in the mucous membrane, fixed with the bone below and encloses a pulp cavity. The crown projects into the pharygeal cavity. Pharynx leads posteriorly into the oesophagus.

II. Progressive zone

It includes the parts of gut where food progresses futher. It has two parts, the oesophagus and the stomach.

The *oesophagus* is a short narrow tube-like structure. Its mucous lining forms seven prominent longitudinal folds, and several smaller folds in between them. A pneumatic duct arises from the dorsal side of the oesophagus and opens into the air bladder. The mucous folds of oesophagus converge posteriorly to form an oesophageal sphincter to check the entry of water into the intestine and also to prevent regurgitation of food. The oesophagus leads into the intestinal bulb.

Intestinal bulb is a long thick walled wide structure extending below the air bladder. The intestinal bulb has an anterior broader cardiac part, and a posterior narrower pyloric part without pyloric caeca which are common in other teleost fishes. The lining of the cardiac part gives honey comb like appearance due to anastomoses of its folds. Similarly the inner lining of pyloric part contains longitudinal folds. The cardiac part receives bile duct and pancreatic duct and a papilla situated just behind the oesophageal sphincter in its roof.

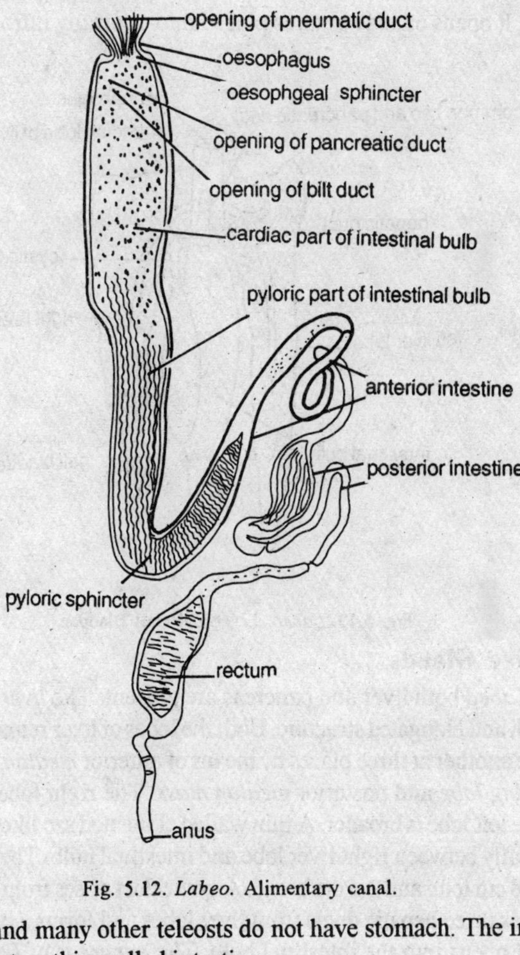

Fig. 5.12. *Labeo.* Alimentary canal.

Labeo and many other teleosts do not have stomach. The intestinal bulb opens into a thin walled intestine.

III. Degressive Zone

It is the part of gut where food is properly hydrolysed and absorbed. It has a very long (about 28 feet) coiled *ileum* which is of uniform diameter. Its internal lining thrown into longitdinal folds, particularly in the posterior part. In the anterior part the folds are transversely displayed. Pyloric caeca, intestinal villi and scroll valve are absent.

IV. Egressive zone

The part of gut that stores undigested wastes for a short period. *Rectum* is wider than intestine and is about 3 feet long. *Rectal gland* is lacking. It opens outside by anus placed mid-ventrally infront of anal fin.

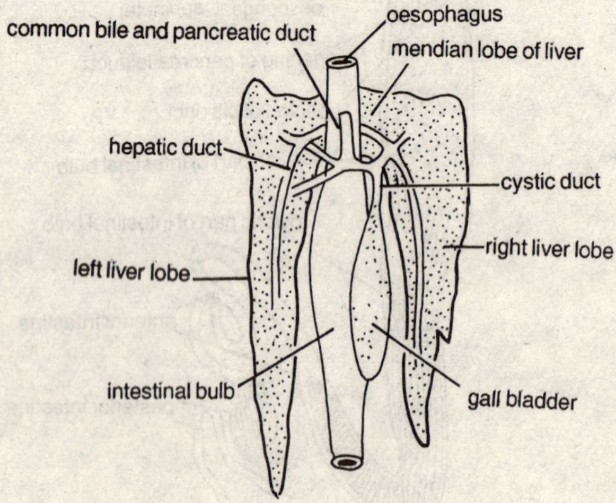

Fig. 5.13. *Labeo.* Liver and Gall Bladder.

Digestive Glands

In *Labeo* both liver and pancreas are present. The *liver* is a bilobed, brownish and elongated structure. Both the lobes of liver remain connected with one another at three places by means of anterior *median lobe*, middle *connecting lobe* and posterior *median mass.* The right lobe is narrower while the left lobe is broader. A thin walled elongated sac like *gall bladder* lies dorsally between right liver lobe and intestinal bulb. The gall bladder is about 8 cm long and 2.5 cm broad. A cystic duct arises from gall bladder. It receives three hepatic ducts from liver lobes and forms a bile duct. The bile duct opens into the intestinal bulb. The *pancreas* of *Labeo* is not a distinct organ. It is in the form of diffused pancreatic tissue scattered in the substance of liver, fat of intestinal mesentery and spleen. The pancreatic duct emerges from the anterior part of the left liver lobe and runs parallel to the bile duct and finally opens into the cardiac region of the intestinal bulb. Both pancreatic duct and bile duct remain enclosed by a common sheath, but remain saparate throughout. Both bile and pancreatic ducts open separately in the intestinal bulb.

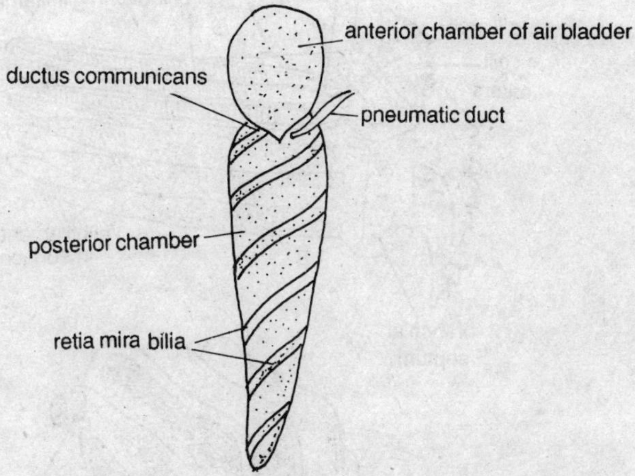

Fig. 5.14. *Lebio*. Air bladder.

Air Bladder

It is also called the *swim blabber.* It is present above the intestinal bulb but ventral to the vertebral column. It is a white, shining, thin walled sac present outside the coelom. It is differentiated into two parts ; small *anterior chamber* and large *posterior chamber* by means of a constriction. A pneumatic duct joins the air bladder and oesophagus together. The fishes with pneumatic duct are called *physostomous.* The air bladder is filled with gas which is a mixture of oxygen, nitrogen and carbon dioxide. It functions like a hydrostatic organ. It regulates the specific gravity of body to make possible the stationary position without muscular efforts.

RESPIRATORY SYSTEM

In *Labeo* the respiratory system can be studied under two heads

(i) Respiratory Tract

It consists of mouth, buccal cavity, pharynx. The mouth is provided with fleshy lips. The buccal region has a pair of *oral valves* that allow water to enter but not to leave it. There are five pairs of *gill-slits* in the pharyngeal wall. *Interbranchial septa* are developed to keep the gill-slits separate. *Gill rackers* protect the gills from hard particles, which do not allow the food particles to pass through the gills. The operculum encloses externally the branchial chamber that opens outside by a large branchial aperture.

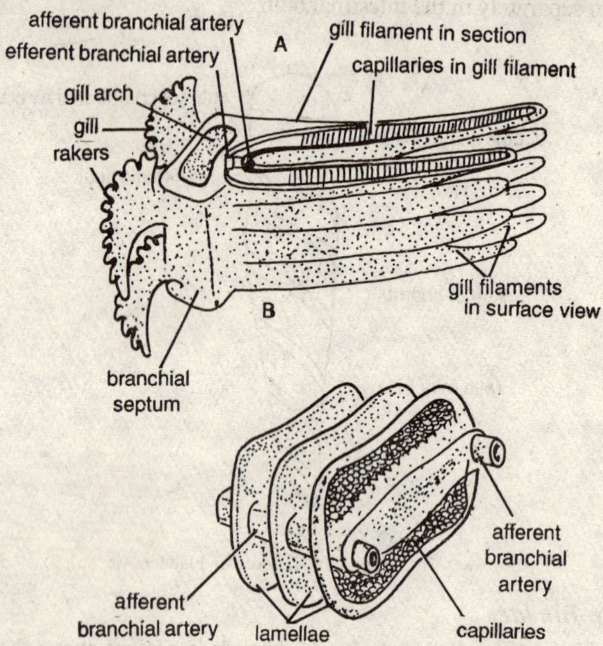

Fig. 5.15. *Lebio*. Showing Gills.

(ii) Respiratory Organs or Gills

There are four pairs of *filiform gills*. The hyoid and the fifth branchial arches are abranch i.e. without gill. Each gill consists of two rows of

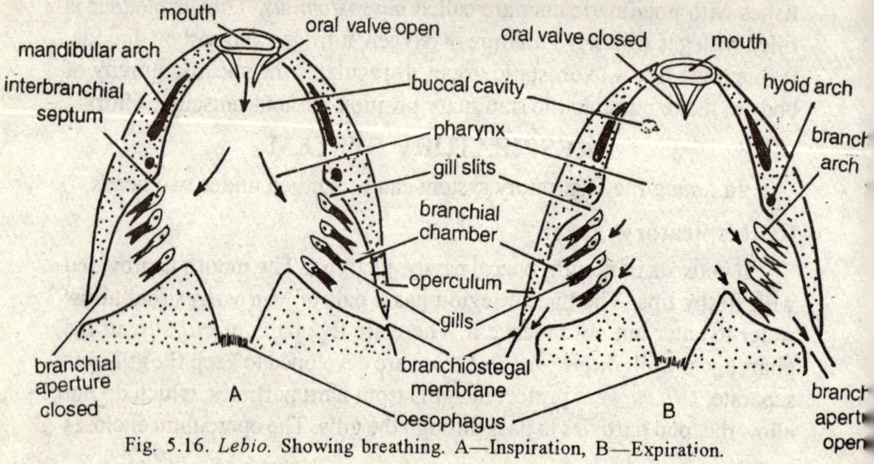

Fig. 5.16. *Lebio*. Showing breathing. A—Inspiration, B—Expiration.

slender processes. The *gill filaments* present along the anterior as well as posterior margin of the interbranchial septum. The gills of *Labeo* are *holobranchs* or complete. The interbranchial septum is reduced and both the gill filaments hang out freely upto the branchial chambers. Afferent branchial arteries supply blood to the gill filament for oxygenation. The efferent branchial arteries leave the gill and carry oxygenated blood.

CIRCULATORY SYSTEM

This system includes *heart, blood vessels* and *blood.*

The *heart* lies in a *pericardial cavity* just beneath the pharynx. The pericardial cavity is separated from the perivisceral cavity by a vertical partition called the *septum transversum. Pericardial fluid,* fills the pericardial cavity. This fluid makes the heart frictionless freely beating and secondly protects it from mechanical injury. The heart is reddish brown organ. It is bichambered and made up of one *auricle* and one *ventricle.* There is no conus arteriosus. A thin walled large *sinus venosus* is present and bears a pair of lateral appendages characteristic of the family. The sinus venosus receive deoxygenated blood from various parts of the body. The sinus venosus pours its deoxygenated blood into atrium through *sinu-atrial pore* provided with *sinu-auricular valve* to prevent the opposite flow of blood. *Auriculo-ventricular valve* to cheek the entry of blood from ventricle to the auricle is present between auricle and ventricle. The auricle lies infront of the sinus venosus. It is oval shaped and is poorly muscularized. The ventricle is the largest and well muscularized structure. It is some what pear shaped. Its inner wall develops muscular ridges called *columnae carneae.* The ventricle opens into a flask-like non-muscular *bulbous arteriosus* through venitrobulbar openining, guided by two *semilunar valves.*

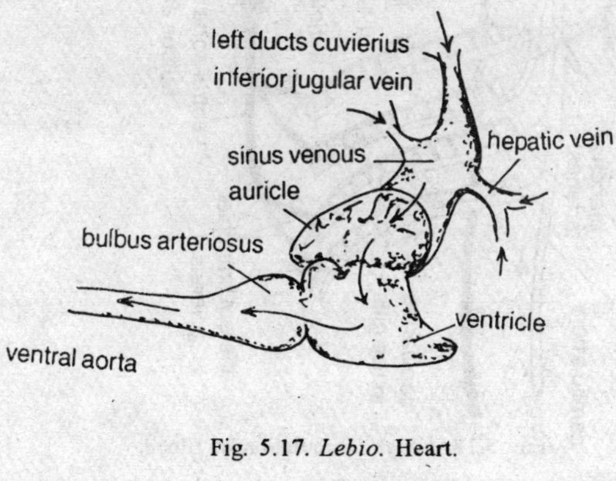

Fig. 5.17. *Lebio.* Heart.

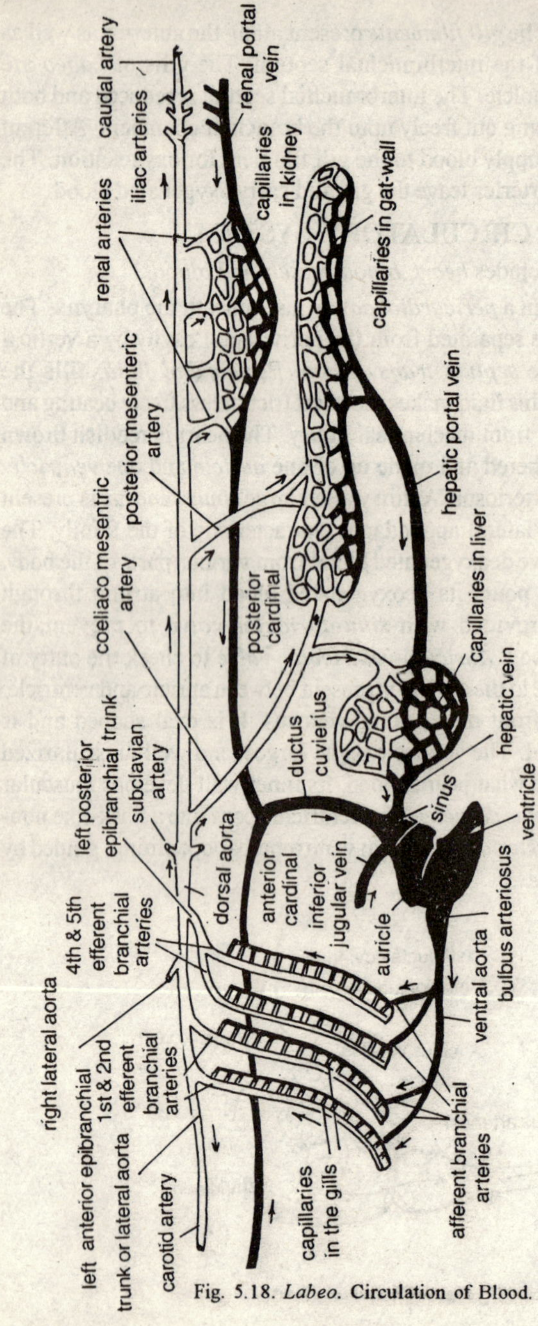

Fig. 5.18. *Labeo*. Circulation of Blood.

In *Labeo* the circulation is single i.e. blood leaves the heart once before given to body organs in one complete circulation. Four afferent vessels carry deoxygenated blood from heart to the gills for oxygenation and thus after another set of four efferent vessel carries oxygenated blood to the dorsal aorta for further distribution. The important blood vessels taking blood from dorsal aorta to body organs are subclavians, coeliac, segmented hypogastric and renal arteries.

Blood is collected from various body organs by means of veins. The blood is finally poured into the sinus venosus by means of three cardinal veins-*two anterior cardinal veins* and a *posterior cardinal vein. Hepatic portal system* is present in *Labeo* but there is no renal portal system.

In *Labeo* the blood is pale and scanty as compared with that of terrestrial forms. It has all the three types of blood corpuscles including *erythrocytes, leucocytes* and *thrombocytes*. Plasma is a watery fluid and is colourless. Haemopoiesis takes place in spleen and kidney.

NERVOUS SYSTEM

Like other verterbrates *brain* lies in the cranial cavity that contains gelatinous fluid, to protect the brain from external injuries. The brain of *Labeo* resembles that of *Scoliodon*. There is a pair of olfactory bulbs enclosing rhinocoel. The olfactory bulbs compensate the absence of olfactory lobes. There is a small cerebrum that remains undivided dorsally. The floor is thickened to form prominent corpora striata. The *diencephalon* is much reduced. It remains covered dorsally by, the optic lobes. The pineal gland arises from the dorsal side of diencephalon. On the ventral side of diencephalon there lies the pituitary gland that functions as an important endocrine gland. Just behind the pituitary there is an oval sac the *saccus vasculosus* which is perhaps a pressure receptors.

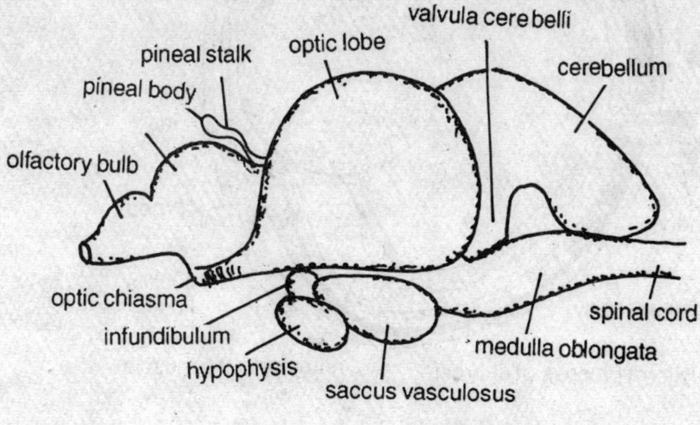

Fig. 5.19. *Labeo*. Brain.

In *Labeo* the optic lobes are very large and form the most prominent part of the brain. The optic lobes are associated with learning, associations, and the performance of relatively complex acts of behaviour. The optic nerves cross and fuse at the crossing to form true optic chiasma. The cerebellum is very large, more or less spherical in shape. Its anterior part is pushed forwards under the optic lobes to form the *valvule cerebelli.* Medulla oblongata is relatively short. It extends back upto the vertebral column in the form of spinal cord. There are 10 pairs of cranial nerves.

SENSE ORGANS

Eyes

Each eye consists of three coats - the *fibrous tunic, uvea* and *retina.* The fibrous tunic forms a complete and very firm layer. Its posterior opaque and whitish part is the sclerotic and anterior transparent cornea which is flat.

The greater part of uvea lines the sclerotic and is composed of soft connective tissue and abundant pigment cells thereby forming *choroid.* At the junction of sclerotic and cornea, the uvea forms *iris* with a circular *pupil.* A layer or *argentia of iris* covers the outer surface of iris. Ciliary body and ciliary muscles are absent.

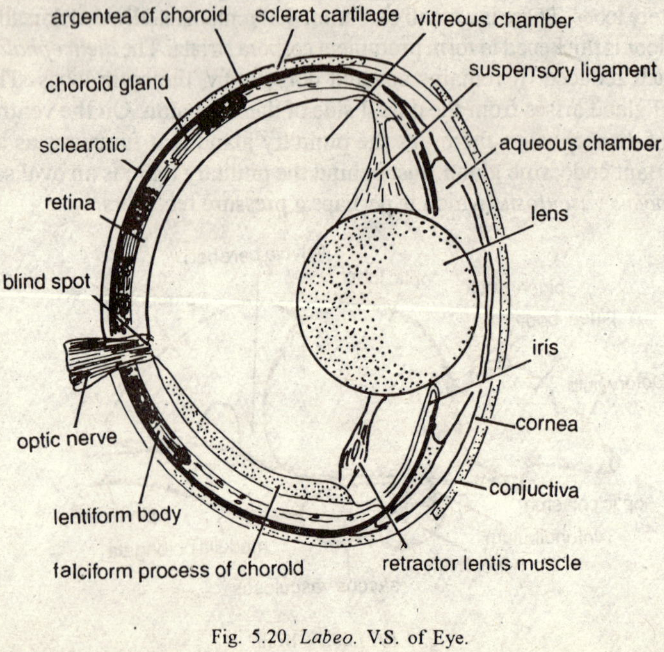

Fig. 5.20. *Labeo.* V.S. of Eye.

The retina is very delicate and lines the whole of uvea. The part of retina from where the optic nerve starts is termed the *blind spot*. Surrounding the entrance of the optic nerves, there is a mass of blood vessels in the choroid called *choroid gland*. A vascular strand, *falciform process*, extends from the point of entrance of optic nerve to the lens where it enlarges as knob the *campanula Halleri or retractor lentis,* which helps in focussing the lens of the eye at distant or nearby objects. *Aqueous* and *viterous humours* fill their gaps in a usual ways.

Ears

The ear is also typical in nature. It is formed of three *semicircular canals* with *ampullae, utriculus, sacculus* and *lagena*. The endolymph contains calcareous particles forming *otoliths*. There are three otoliths in *Labeo*. At the junction of utriculus and sacculus, the *endolymphatic duct* originates to maintain the position of membranous labyrinth. The membraneous labyrinth is filled with *endolymph* and the space surrounding it is filled with *Perilymph.* Sensory spots in the form of *cirstae* and *maculae* are also found. Sacculus is well developed where as lagena is rudimentary.

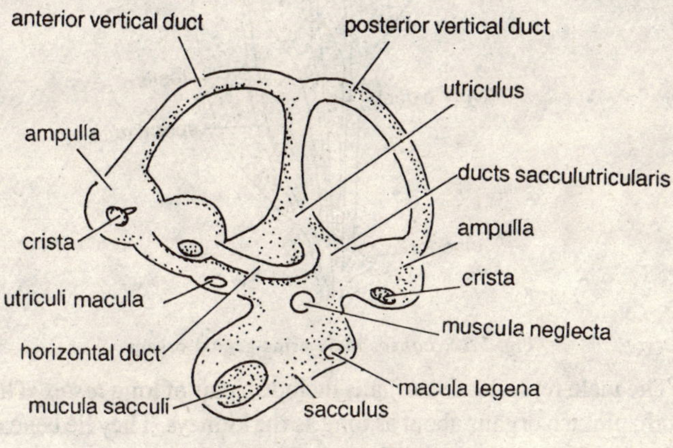

Fig. 5.21. *Labeo.* Ear.

URINOGENITAL SYSTEM

In fishes both urinary and reproductive organs are intimately associated together but in *Labeo* they are quite independent. The sexes are also separate.

The kidneys are of *opisthonephric* type and extend as long brownish structures along the length of the body cavity below the vertebral column and above the gut. Their free anterior ends have knob like *head kidneys* which are the non-functional embryonic pronephric kidneys retained in adults. The kidneys are dilated in their middle region and tapers posteriorly. Their broad middle and narrow posterior parts are united and form the functional renal part containing *uriniferous tubules.* To carry nitrogenous wastes there is a pair of *ureters* running along the length of their respective kidneys. Both the ureters unite into a common urinary duct within a thin walled sac like *urinary bladder* that opens outside through the urinary aperture situated behind the anus.

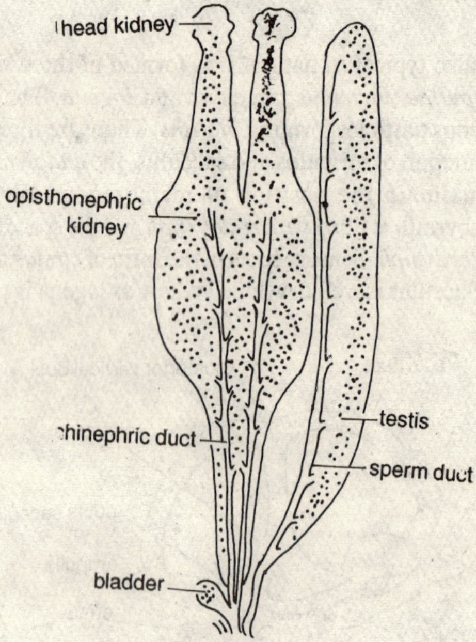

Fig. 5.22. *Labeo.* Male urino-genital organs.

The male reproductive organs include a pair of long *testes.* They are smooth, pinkish organs about as long as the kidneys. They lie beneath the air bladder and are suspended from the body wall by folds of peritoneum, called *mesorchia.* The germ cells of testes undergo spermatogenesis and produce sperms. There is a pair of *sperm ducts* communicated with their respective testis by means of several fine ductules it receives sperms. The sperm ducts join posteriorly and open out by the *genital aperture.*

In female there is a pair of *ovaries.* They are long organ extending nearly the entire length of the abdominal cavity beneath the air bladder. The ovaries remain suspended in the abdominal cavity by the *mesovarium.* During breeding season the germinal epithelium of ovary undergoes oogenesis and produces large number of ova and thus the ovary gets folded. *Labeo* is oviparous and the female lays thousands of relatively small and yolky eggs. There is a pair of *oviducts* for the purpose. The oviducts are formed by the peritoneum that surrounds the ovaries. The two oviducts unite and open to the exterior by the *genital pore*, lying between the anus and urinary aperture.

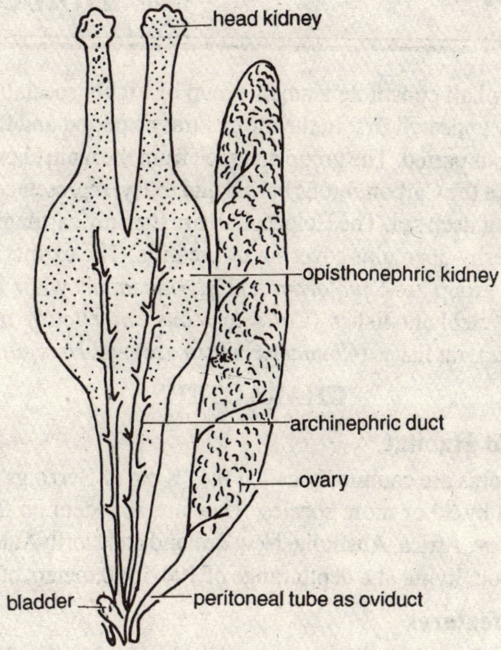

Fig. 5.23. *Labeo*. Female urino-genital organs.

During July and August *Labeo* lays eggs in water. About 1,905,000 eggs are laid at a time. Fertilization takes place in water. The development is direct i.e. without larval form. The newly hatched young ones are called *fry*. They start feeding and grow into fingerlings and become sexually mature in about two years.

Revision Questions

1. Describe the habits, habitat and external features of *Labeo rohita.*
2. Describe the respiratory organs and mode of respiration in *Lebeo.*
3. Describe the digestive system and mode of feeding in *Labeo.*
4. Write short notes in : (a) Operculum, (b) Air bladder, (c) Gill rackers.

6

Holocephali

The Holocephali constitute a small group of highly specialized marine fishes. They appeared first in the lower Jurassic period and flourished in the Cretaceous period. The group diverged from the main chondrichthyan stock early in the Carboniferous Period and today, represented by strange Chimaeras of deep sea. The Holocephali are the only cartilaginous fishes having a fleshy opercular covering of the gills. The extinct representatives of the group are *Squaloraja* and *Myriacanthus* while the existing species are elephant fishes (*Callorhynchus antarticus),* rabbit fishes (*Hydrolagus*), rat fishes (*Chimaera monstrosa*) and *Harriotta.*

CHARACTERS

Habits and Habitat

Chimaeras are commonly called the "King of Herrings". These are represented by 30 or more species. They are abundent on the coasts of Japan, Europe, Africa, Australia, New Zealand and North America. These are benthonic, living at a depth range of 100-1500 meters of the sea.

External features

Body is shark-like in appearence. The head is large and compressed with a small mouth and the body tapers to the tip of a long and slender tail. There is a single *nasal aperture* and four pairs of *gills*. The gills are covered by a fold of skin forms *operculum* on each side. They open out side by a single *branchial aperture*. The eyes are very large. The spiracle and cloaca are absent. The urinogenital aperture being distinct from anus. There are two *dorsal* and a *ventral fins*. The tail is *heterocercal*, but in *Chimaera* it is produced into a whip-like filament. The *pectoral fins* are large but *pelvic fins* are small. Skin is gererally smooth that is without scales and silvery. However, in *Chimaera* the *placoid scales* are restricted to certain parts of the body. The lateral line system is grooved in *Chimaera* but a closed tube in *Callorhynchus.*

In the male, a shallow glandular pouch is present in front of the pelvic fins and contains a peculiar anterior clasper covered with denticles. The function of clasper is unknown. In addition, a pair of ordinary claspers are also present behind the pelvic fins and a frontal clasper on the dorsal surface of head.

Endoskeleton

The endoskeleton is cartilaginous. The vertebral column consists of a persistant notochord with cartilaginous arches. Calcified rings are embedded in the notochordal sheath in *Chimaera.* The neural arches of the first few vertebral are fused to form a plate giving attachment to the first dorsal fin.

The skull is *holostylic.* The palato-quadrate is fused with the neurocranium, providing support for the lower jaw. Mouth has crushing plates, two pairs in the upper jaw and one pair in lower jaw. The plates are without enamel but vitrodentine and reduced pulp cavities are present. Three cartilaginous rods grow forwards from the cranium to support the snout in *Callorhynchus.* In *Chimaera* these rods are very much reduced. The pituitary fossa is deep and inclined backward. The occipital condyle is saddle-shaped articulating with vertebral column. The hyoid resembles the branchial arches. Above the epihyal there is a small pharyngohyal. Two halves of pelvic girdle are not fused, Each half consist of a narrow iliac region and a broad pubo-ischial region.

Digestive System

The mouth is small. It ia bounded by three lip-like folds and equipped with the tooth plates. The plates have irregular surface and sharp cutting edges. There are two pairs of plates in the upper jaw and one pair in lower jaw. Vomarine, palatine and mandibular teeth are present.

The alimentary canal is straight and a simple tube. The oesophagus is short directly opening into intestine. The stomach is absent. Intestine has a well developed *spiral valve*. It leads into rectum. The rectum is short and opens to exterior by an anal aperture.

Respiratory System

The branchial organisation of holocephalian differs from those of sharks and bony fishes. A mandibular pseudobranch is absent. The respiratory organs consist of three pairs of *holobranchs* (complete gills) on the first three branchial arches and two pairs of *hemibranches* on the hyoid arch and the fourth branchial arch. The fifth arch is *a branch*. The *interbranchial septum* is reduced.

Circulatory System

It resembles that of *Scoliodon* in all essential respects. The heart consists of *sinus venosus, atrium, ventricle* and *conus arteriosus.* The conus arteriosus has three rows of valves.

Nervous System

The cerebral hemispheres are small in size. They are spindls- shaped and connected to olfactory bulbs by means of narrow stalks, The diencephalon is long, thin and trough shaped. The pineal body is small and rounded and has a hollow stalk. The pituitary body has intra- and extra-cranial portions. Cerebellum is small. Medulla oblongata is produced laterally into restiform bodies.

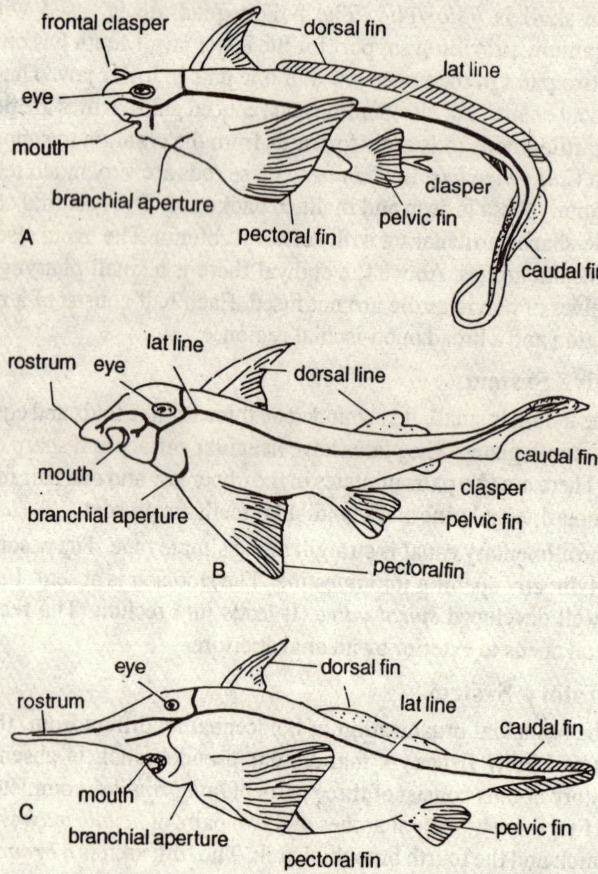

Fig. 6.1. A—*Chimaera monstrosa,* (B) *Callorhynchus antarticus,* (c) *Harriotta sp.*

Sensory organs

The *eyes* are large in relation to body, because they live in the total darkness of the deep sea. The *internal ear* consists of three *semicircular canals*. The vertical canal forms the crus, also receves the horizontal canal. The *ductus endolymphaticus* opens outside by pore.

The lateral line canals are open and are well developed on the head and on the under side of snout.

Urinogenital System

The excretory organs are a pair of lobed deep-red *kidneys*. These are short and stont. They are slightly longer in males. Large number of uriniferous tubules are present in each kidney. The kidnay is segmented both anteriorly and posteriorly. The porterior part is narrow. The female reproductive organs consist of a pair of *ovaries*, *shell glands*, and *uteri*. In the male there are two ovoid *testes*. A number of *vasa efferentia* form a *vas deferens* which comes out from each testis. Near testis the vas deferens forms a coiled structure called *epididymis*. The epididymis is closely applied to the anterior part of the kidney. The sperms become mature in the epididymis and form *spermatophores*. The terminal part of vas deferens is swollen to form *vesicula seminalis*. The later is divided into many chambers by transverse septa. The spermatophores are stored in these chambers and are finally passed into the urinogenital sinus. The fertilization is internal. The animals are oviparous. The eggs become surrounded by a horny egg shell secreted by shell glands.

AFFINITIES OF HOLOCEPHALI

Holocephalians occupy a position between cartilaginous (elasmobranchs) fishes and bony (teleosts) fishes. Besides they retain cartain primitive as well as a number of specialized characters which are not present in cartilaginous and bony fishes that is why this group has been given the status of independent subclass.

Resemblances with the Elasmobranchs

1. A cartilaginous endoskeleton is present. The replacing and membrane bones are absent.
2. The vertebral column has a persistant notochord with cartilaginous arches.
3. The limbs and girdles are built on the elasmobranch plan.
4. Paired fins are present and the tail is heterocercal.
5. The skin is silvery and smooth. The placoid scales develop in the form of denticles during the embryonic life of chimaeras. In adults they are lost but a few are present on the claspers.

6. A pair of claspers are present posterior to the pelvic fins in males, as in elasmobranchs. Besides these, anterior and frontal claspers are also present.
7. The basic plan of brain resembles with that of sharks, but the greater development of restiform bodies and elongation of diencephalon show remarkable differences.
8. The air bladder is absent.
9. A spiral valve is present in intestine.
10. The conus arteriosus is present in heart bearing three rows of valves. The urea is retained in blood for osmoregulation.
11. The nasal and labial cartilages along with the orbital margins are present.
12. The lateral line canals are distinct and suggestive of the shark.
13. The eggs are essentially similar to those of oviparous sharks and encased in horny capsules.
14. A posterior separate superficial opthalmic foramen is present.

Resemblances with Teleosts

1. Presence of an operculum and a single opercular opening to the exterior.
2. The gill-filaments project beyond the interbranchial septum. The interbranchial septa of the gills are reduced and are not continuous with the skin.
3. The spiracle is absent.
4. Four complete aortic arches serving the four gill openings.
5. The urinogenital aperture lies behind the anus so that there is no cloaca.

Primitive Characteristics

Following primitive characters are seen in Holocephalians :

1. The cartilaginous endoskeleton such as the persistent notochord.
2. For each gill-slit, separate external opening is present.
3. Large opercular covering of the gill-slits.

Specialized features

In addition to above mentioned characters, the Holocephali have following specialized characters

1. The jaw suspension is *holostylic.* The palatoquadrate is fused with cranium. Neither the skull nor the jaws are attached to the hyoid

arch. From such kind of jaw suspension the group derives the name Holocephali (*holo* =whole; *caphalos* = head).

2. The dentition is peculiar in having large plates. Teeth are not covered by enamel but by layer of vasodentine. The pulp-cavity is reduced, sharp edges are there in the plates.
3. Presence of unusual claspers (anterior and cephalic claspers).
4. Absence of scales, spiracles and cloaca.

Thus group Holocephali have both, primitive as well as specialized characteristics. This have led most authorities to believe that it forms an aberrant group of fishes. The Holocephali arose from some shark ancestor during the Mesozoic Era, possibly in Triassic period. Originally they were considered as an order under subclass - Bradiodonti of class Elasmobranchii but now it has been given the status of independent subclass.

7

Dipnoi

The Dipnoi are commonly known as 'Lung-fishes'. It is the most interesting group of bony fishes and shows close affinity to the teleosts on one hand, and to the amphibians on the other. They evolved during the Devonian period and flourished during the Permian and Triassic *Periods*. Several fossil forms such as : *Dipterus* (Dipteridae), *Dipnorhynchus, Phaneropleuron, Uronemus* (Uronemidae) and *Ceratodus* (Ceratodontidae) are on record. At present, the Dipnoi are represented by three genera, occuring widely separated tropical and subtropical fresh water habitats. These are *Neoceratodus, Protopterus* (Protopteridae) and *Lepidosiren* (Lepidosirenidae). *Neoceratodus* occurs in south eastern Queensland (Australia), *Protopterus* in Africa and *Lepidosiren* is South America. There are four species of *Protopterus—P. annectans, P. aethiopicus, P. dollei* and *P. amphibius.*

Class Dipnoi comprises of only two orders Monopneumona (with single lung) e.g. *Neoceratodus*; Dipneumona (with two lungs) e.g. *Protopterus* and *Lepidosiren.*

External characters

Body somewhat eel-like in appearance ranging in length from less than a feet to nearly six feet in different species. Paired fins are more or less acutely lobate and overlapping cycloid scales are embedded in connective tissue of dermis. In the extinct from, *Dipterus,* the scales were thick and were covered with a layer of *cosmine*. Paired fins are supported by dermal fin rays which are jointed, branched and bony in nature. The fins are of actinopterygian type. The tail is *diphycercal* with continuous dorsal and ventral median fins. *Dipterus* had two separate dorsal fins and one anal fin and a *heterocercal* tail.

Snout bears a pair of *nostrils* on the ventral surface. These nostrils open into the mouth by means of a pair of internal nostrils, facilitating aerial respiration as in frog.

Fig. 7.1. Fossil Lungfishes. A—*Dipterus* (Devonian), B—*Phanero-pleuron* (upper Devonian)

The mouth is situated on the ventral surface of the head whereas cloacal aperture lies at the root of the tail. Spiracles are absent and there is only one external branchial aperture due to presence of an operculum.

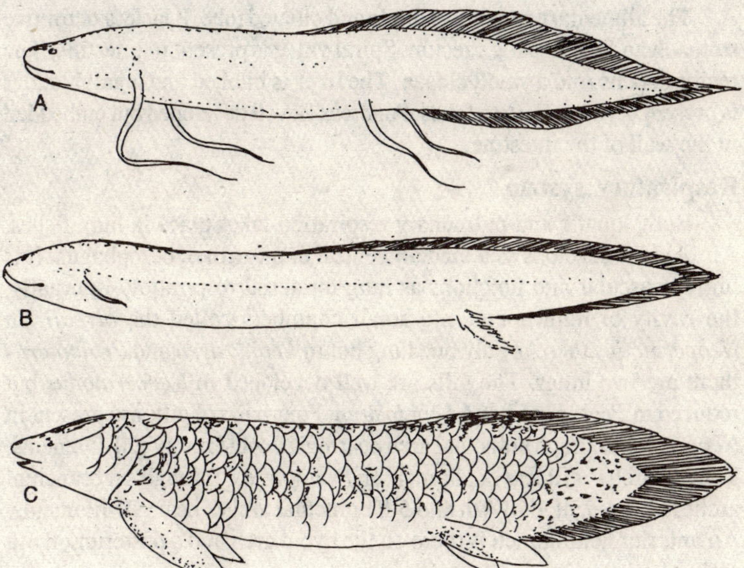

Fig. 7.2. Living Lungfishes. A—African Lungfish (*Protopterus*), B—South American Lungfish (*Lepidosiren*), C—Australian Lungfish (*Neoceratodus*).

Endoskeleton

The skeleton is largely *cartilaginous*. There is persistant and unconstricted notochord covered with a fibrous sheath. The vertebral column consists of paired basidorsal and basiventral cartilagenous elements only. The former form neural arches and neural spines while later form haemal arches and haemal spines. Pleural ribs are present in body wall.

In the skull premaxillae, maxillae and nasal are absent. The hyomandibulars are also absent. Dentaries are absent in lower jaw and the teeth are present on coronoid. Large dental plates are firmly fused with jaw bones and are not replaceable. They are composed of thick dentine round numerous branching pulp cavities. The skull is *autostylic* i.e. hyoid arch does not take part in the formation of suspensorium. The palatoquadrate bar is firmly fused with the cranium infront and behind. Exoccipitals and palatoquadrate cartilages are present. The hyoid and the branchial arches are cartilaginous.

The fin skeleton consists of a central axis of small cartilaginous rods with radials on both the preaxial and postaxial sides.

Digestive system

The alimentary canal is a simple and ciliated tube. It lacks a definitive stomach and the hepatic caecum. Spiral valve is present in intestine. The rectum opens into a small cloaca. The liver is bilobed and a gall bladder is present between the two lobes. Pancreas is well developed but embedded in the wall of the intestine.

Respiratory system

Both aquatic and pulmonary respiration takes place in lung fishes. Air bladder develops as a median ventral outgrowth of oesophagus. It is highly vascular and functions as lung for aerial respiration. Internally, the cavity of lung opens into small chambers called the *alveoli*. In *Neoceratodus* there is only one lung but in *Lepidosiren* and *Protopterus* there are two lungs. The gills are well developed in *Neoceratodus* but reduced in *Protopterus* and *Lepidosiren*. Four pairs of gills are present in *Neoceratodus.* Each branchial arch bearing a double row of gill-filaments. The 5th arch is gill less or abranch. In *Protopterus*, through six branchial arches are present, only 4th and 5th branchial arches bear gill-filaments. An anterior hemibranch is there on the hyoid arch and a posterior on 6th arch.

Circulatory system

The heart is three chambered, consisting of two *auricles* and a *ventricle*. The *sinus venosus* is imperfectly divided into two chambers. The deoxygenated blood runs through the right side of the heart and the oxygenated blood through the left side. The auriculo-ventricular opening is regulated by means of a fibrous plug. The oxygenated blood is brought to the left auricle by pulmonary veins from the lungs. The deoxygenated blood from rest of the body is poured into the right side of the sinus venosus and pass into right auricle. The ventricle leads into the conus. The conus is divided into two channels by means of a valves. The arrangement of valves is such that the oxygenated blood flows through the first two arches into carotid and cerebral circulation. Venous blood tends to go into the posterior branchial arches and is finally carried to the lungs.

The ventral aorta is small and gives off four pairs of afferent branchial arteries. There is no distinction of afferent and efferent branchial vessels in abranch arches of *Protopterus.* In the arches that bearing gills, paired efferent branchial vessels are present. These unite to form the dorsal aorta. From paired dorsal aortae, pulmonary arteries arise and carry blood to lungs.

Two precaval are present which receive blood through inferior jugular, anterior cardinal and subclavian veins. The posterior cardinal vessle directly opens into the sinus venosus forming the post caval. The post caval is absent in fishes but occurs in tetrapods. The caudal vein divides into right and left renal portal veins and enter the kidneys.

Nervous system

The cerebral hemispheres are well developed in brain and are quite large. A single optic lobe is present. The pineal body is also present. The cerebellum is small.

Urinogenital system

The excretory system consists of a pair of mesonephric kidneys. They discharge the excretory product into cloaca through the ureters.

A pair of elongates testes are present in male. The anterior part serves for the spermatogenesis and the posterior part acts as seminal vesicle. Both the vasa deferentia unite to open into cloaca. Mullarian ducts are present in vestigial form.

In female, a pair of ovaries are present near the kidney. Oviducts open anteriorly into the coelom and posteriorly unite to open into cloaca.

AFFINITIES OF DIPNOI

Characters of Dipnoi show that this group has not undergone much changes during the long period of existence Dipnoan constitute a remarkable group of fishes, that show affinities with the long fishes as well as amphibians but finally they have been given the status of a separate class because of presence of several striking features.

Resemblance with Actinopterygians

1. Both existed from the mid Divonian period and represented by a few extinct genera.
2. A pair of swim bladder connect with the parynx on its ventral side.
3. A spiral value present in intestine resemble with that of *Polypterus.*

Resemblance with Crossopterygians

1. Both are similar in appearence
2. Like Dipnoans the Rhipidistians (earlier crossopterygian) also had enamaloid cosmoid scales. Coelocanth, however, have a slightly different type of cosmoid scales.
3. Paired fins are lobate and of archipterygian type.
4. Degenerated gut diverticulum of coelocanths corresponds to respiratory swim bladder of lung fishes.
5. Intestine has a long spiral valve.
6. The two dorsal fins of extinct *Dipterus* resemble the similar fins of rhipidistians.

Resemblance with Amphibia

1. Dipnoans are capable for living in marshy places. Lungs are present for breathing in addition to the gills.
2. Both, the swim bladder of dipnoans and lungs of amphibians, receive their blood supply from the sixth aortic arch.
3. The heart is three chambered i.e. having two auricles and a ventricle. The conus is divided into two channels by a longitudinal valve.
4. Separate pulmonary artery and pulmonary veins are present in adult to separate pulmonary and systemic circulation respectively. A renal portal vein and an inferior vena canae are present.
5. Larvae of *Lepidosiren* and *Protopterus* possess external gills and can be compared to the tadpole larva of frog.
6. One pair of external nostrils are present and open into the buccal cavity through internal nostrils.
7. Jaw suspension is autostylic.
8. The pattern of urinogenital system is similar except that in male dipnoans where the Mullerian ducts also develop.
9. Various embryological similarities include holoblastic cleavage, formation of yolk-plug and possession of suckers and external gills by larvae of *Lepidosiren* and *Protopterus.*

Thus Dipnoi constitute a remarkable group of those animals which although fish-like in many characters but possess a number of amphibian-like characters . Besides they possess a number of specialized characters not present in fishes as well as in amphibians. Only a few resemblances are supposed to be due to convergent evolution. According to some workers, the Dipnoi are the direct ancestors of amphibians. But this view is not accepted by all the ichthyologists. Dipnoi show structural resemblances with Placoderms, Elasmobranchs and Amphibians and the Crossopterygii. They have a close affinity with the Crossopterygii. According to *Romer* lung fishes and the Rhipidistians (Crossopterygii) are clearly derived from a common ancestor.

Modern Dipnoi, thus appear to be a group that has not undergone much changes during long period of their existence after the Permian period. It is true that Dipnoi are not the ancestors of Amphibia but the ancestors of these two groups were possibly derived from some common stock. Thus Berg's view of giving the Dipnoi the status of an independent class is justified.

8

Luminous Organs

In common with many other animals of similar habitate photophores or luminous organs are characteristic structures in many deep sea fishes and bathypelagic fishes living in depth of 500-2000 meters. These organs are probably specialised gland cells of the epidermis and show considerable variations in their number and mode of distribution on the body. Generally they occur along the lateral and ventral sides of the body and the head. They may be arranged in one or two rows extending on the sides from head to tail as in *Holosauropsis* or may be present on certain parts of the body only. In the case of *Opostomias*, these organs are arranged in the form of transverse bands on the body.

The light producing mechanism has also been studied in organisms other than fishes. Glow worm or fire fly, *Noctiluca*, certain jelly fishes or medusae, ctenophores and shrimps have light producing mechamism. The production of light provides yet another example of transformation, for the luminous organs or photophores owe their origin to the modification of certain gland cells in the skin. These organs vary ranging from simple aggregation of cells associated with *luminous bacteria,* to elaborate and powerful structure with lens and reflector. Some fishes seem to have the power of emitting light without possessing any definite light organ, but in some forms at least, this is due to the presence of luminous bacteria. A member of the Rat-tails or Grenadiers (*Malacocephalus laevis*) has a complicated gland near the anus in which these bacteria are contained. When the fish is disturbed it secreates the bacterial mass through a duct.

Luminous Organs

In fishes the ability to produce light has reached its perfection. Among the cartilaginous fishes, a few deep sea sharks belonging to the order Squalidae emit light. This has been described as a vivid and greenish phosphorecent gleam, and has been shown to be due to the presence of numerous tiny light organs of very simple structure scattered over the skin. *Elmopterus* of the Atlantic and Mediterranean, known to fisherman as "Darkic Charlie" has been kept alive in the Nepales aquarium and the production of light carefully observed. Another form found near Ceylon was placed on the deck of a ship after capture, and continued to emit a luminous glow until its death three hours later.

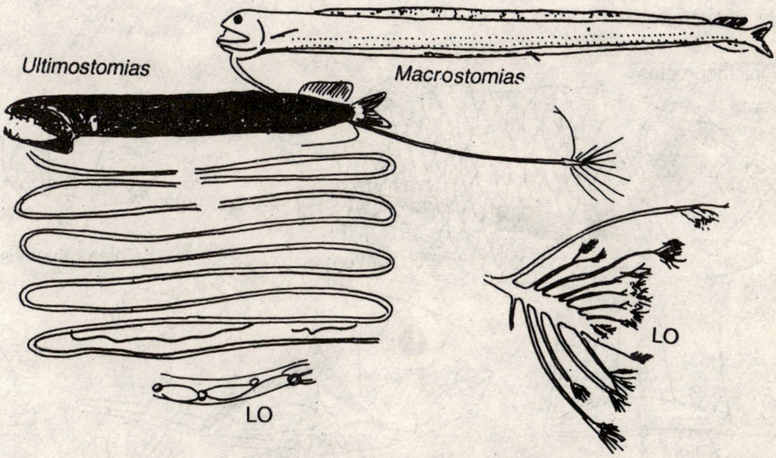

Fig. 8.1. Light Organ in Stomatoid Barbels.

In bony fishes there is a large variety of light organs. One of the well known example is the wide mouths, *Stomiatoidei*, inhabitants of open ocean. There are typically two rows of organs, or photophores, on either side of the fish, one of the belly and another parallel to it and near the lower edge of the side, but in some species additional series may be developed above these. The rows generally extend continuously from behind the head to the tail, but may be interruped in some species and confined to certain region of the body and tail. Each organ is composed of a group of gland cells which are shaped like a-cup. The inside of the cup has a lining of black pigment to form a reflector, similar to that used in a bull's eye lantern. The mouth of the cup has a diaphragm supporting a lens. The gland cells emit light which is reflected through the lens. A still more curious organ is found in a Rare Berycord fish from the Indian Ocean known as *Anomalops*. This fish has a large organ below the eye, placed on a movable flap, so that when light is not wanted it can be turned inwards and received into a cavity underneath the eye. Luminosity in this species is due to masses of symbiotic luminous bacteria living in the cells comprising the organ.

In lantern fishes belonging to Myctophlidae the photophores are fewer in number. They are large and brighter having the appearance of glistening jewels or small mother of pearl buttons. They have a complicated arrangement of short rows and groups which is nevertheless perfectly symmetrical. Many fishes have a special organ above or below the eye and others possess a very powerful head-lamp covering the greater part of the snout.

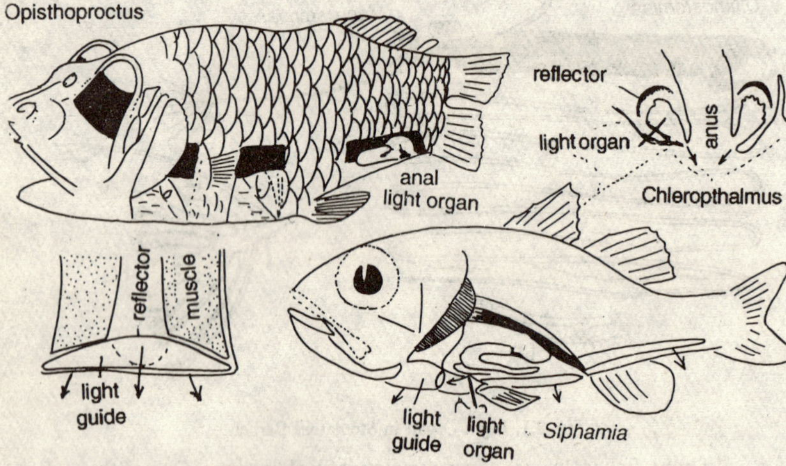

Fig. 8.2. Bacterial light organs.

In the deep sea members of the order Ceratioidei . The bait is luminous and can probably be switched on at will, serving to attract smaller fishes within reach of the Angler's jaws. Another remarkable fish, known as *Ipnops* is blind and has the whole of the upper surface of the flattened head occupied by a pair of large organs lying beneath the transparent superficial bones of the roof of the skull and believed to be luminous.

The luminous fishes may be classified into two groups.

Some species are self luminous but in others the light is produced by symbiotic bacteria. The best known example of the fish with luminous bacteria is the *Malacocephalus* having a large internal luminous gland opens by a duct to the out side just anterior to anus. The gland is richly supplied by the blood capillaries and a network of nerves that lies in the connective tissue capsule of the gland. The gland contains luminous tubules which harbour luminous bacteria. In *Monocentris japonicus* the gland lies on the underside of the tops of the lower jaw, and in

Photoblepharon under the eye. In *Photoblepharon* the bacteria are found in long parallel glandular tubules with abundant blood vessels. The organ has pores opening to the exterior at the anterior end and a reflector layer at the bind end. These organs produce light for long periods due to bacterial luminescence, hence mechanical devices are developed to turn the light on and off. In *Anomalops* the light is cut off by moving the light organs downward by means of a hinge, so that it comes into contact with a black pigmented tissues. In *Photoblepharus*, a fold of black tissue can he drawn up like an eye lid over the organ, that cut off the light.

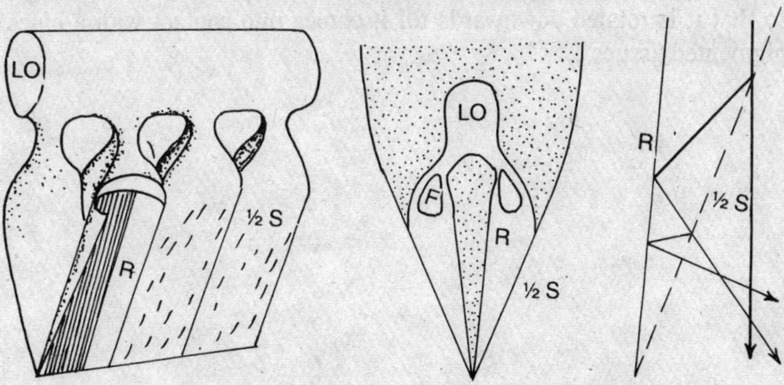

Fig. 8.3. Organisation of Ventral Photophores.

In some elasmobranchs and teleosts self luminous photophores, are found. The simple photophores consists of a series of radially arranged glandular tubules receiving branches from the adjoing cranial or spinal nerves.

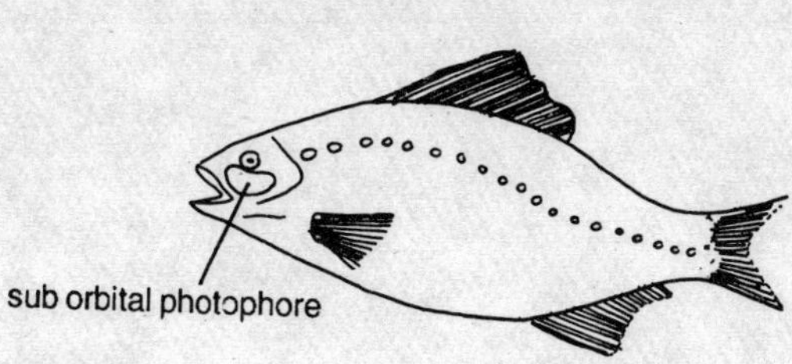

Fig. 8.4. *Photoblepharon* showing photophores.

Control of Photopheres

The process of luminescence is under the nervous and hormonal control. The photophores are inervated properly to activate the photocytes by their efferent fibres. The suprarenal gland may also activate them. *Adrenaline* also show its remarkable effects. To change the brightness certain fishes use mechanical methods. *Idiacanth* is able to pull down a pigmented sheath infront of the light producing surface to conceal the light organ. In *Photoblepharon,* a memberanous black curtain lies along the central edge of the organ, and is pulled up to conceal the light. In *Anomolops* the light organ is provided with a hinge at the anterior edge, so that it is rotated downwards till it comes into contact with a black pigmented tissues.

9

Air Bladder

In great majority of fishes an air bladder lies between the alimentary canal below and, the kidneys and vertebral column above. This bladder is absent in elasmobranchs. The suggestion that the small caecum on the dorsal side of oesophagus in certain sharks, as for example *Acanthias* represents it, has not been confirmed by later workers.

Perhaps the most primitive air bladder is seen in *Polypterus*. The bladder is double. Its right lobe is long and tubular while the left is small and oval. At the anterior end the two lobes fuse into an unpaired chamber which opens in the oesophagus on its ventral side. The organ is devoid of internal sacculations.

The Dipnoi are provided with a very similar muscular vestibule opening into oesophagus on the ventral side. The vestibule is single in *Neoceratodus* and double in *Lepidosiren* and *Protopterus*. The central cavity of bladder communicates with alveoli in the bladder wall which may open into ducts leading to caecal sacculi (*Protopterus*).

In the rest of bony fishes the bladder is a median dorsal diverticulum of alimentary canal. It usually opens into oesophagus by *ductus pneumaticus*. The ductus is short and wide in lower forms like *Amia* and *Lepidosteus* but long and narrow in the higher forms. It occurs in the form of an open duct in the lower bony fishes and it becomes reduced and finally disappears in the specialized groups like Mugiliformes and actinopterygian fishes. On this basis, bony fishes may be divided into *Physostomi* with open duct and *Physoclisti* with closed bladder. Since physoclistous forms occur among physostomous group and open ducts are retained some times in physoclistous group. This classification is not of importance and has been abolished. In *Clupia* in addition to ductus, the bladder opens to outside by a tubular canal from its posterior end. In a physoclistous fish a duct from bladder opens in the right branchial cavity. Since the bladder is a diverticulum of alimentary canal there is a relationship between the layers of alimentary canal and those of swim bladder. The tunica externa is comparable to muscularis externa and tunica interna to mucosa and sub-mucosa.

Layers of alimentary canal	*Layers of air bladder*	
	Description	*Terminology*
1. Serosa	Peritonial covering of the ventral surface of bladder.	Serosa
2. Muscularis externa	Dense connective tissue and non-straiated muscles of longitudinal and transverse types.	Tunica externa
3. Sub-mucosa	Loose layer of fibroelastic tissue.	Sub-mucosa
4. Muscualaris mucosa	Smooth muscle cells	Muscularis
5. Mucosa	Inner flat epithelium	Inner epithelium

The air bladder is not universally present in many bony fishes e.g. flat fishes. Its form is also different in various fishes and is usually correlated

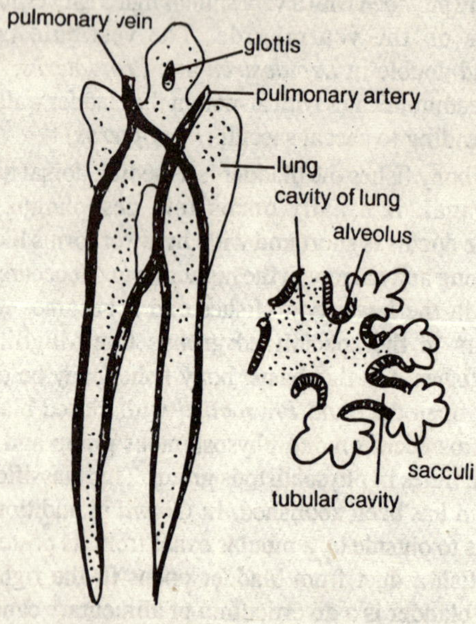

Fig. 9.1. *Protopterus*. Air bladder

with the form of body. It is tubular (*Xenentodon*), fusiform (Broad bodied fishes), ovoid (*Wallago*), heart shaped (short and broad fishes) or horse shoe shaped (*Aiter*). A transverse constriction frequently divides it into two inter-communicating sacs as in Cyprynoid fishes or by two constrictions into three sacs as in *Ophidium*. Not rarely the organ extends into the tail as in *Notopterus* or gets reduced as in *Clarias* which is associated with smallness of body cavity necessiating its prolongation into the tail or getting reduced.

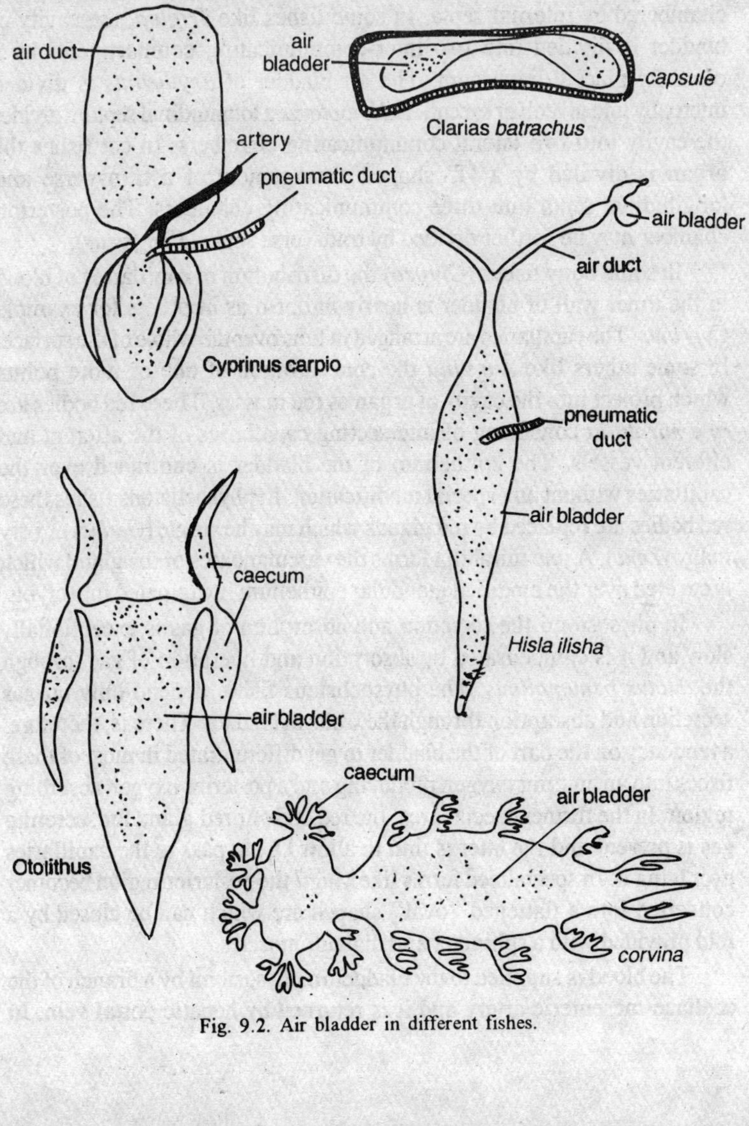

Fig. 9.2. Air bladder in different fishes.

In many bony fishes a complex system of simple or branched caecal out growth come out from bladder as in *Otolithus*. The caecal out growths are given off from the anterolateral angles of bladder. Each dividing into two elongated sacs. In *Callichthys* the tabular branches come out from each side of bladder which is divided and sub-divided extending to mid-dorsal and mid-ventral line where they become continuous with the branches of other side.

In addition to division of bladder externally the organ is often chambered by internal septa. In some fishes like *Triglea*, the cavity of bladder is divided into two inter-communicating compartments by a centrally placed diaphragm. The air bladder of *Erythrinus* is divided internally into alveoli or sacculi. In *Notopterus* a longitudinal septum divides the cavity into two lateral communicating chambers. In cat fishes the organ is divided by a 'T' shaped arrangement of a transverse and longitudinal septa into three communicating chambers. The posterior, chamber may be further divided by transverse septa (*Pangssius*).

In some bony fishes (*Clupea*) the distribution of capillaries of blood in the inner wall of bladder is nearly uniform as in others for example *Cyprinus*. The capillaries are arranged in tufts over the whole of the surface. In some others like *Anguilla* the concentration of one or more points which project into the cavity of organ as red masses. These red bodies are *rete mirabilia* consisting of intersecting capallaries of the afferent and efferent vessels. The epithelium of the bladder is continued over the capillaries without any special modification. In physoclistous fishes these red bodies are replaced by red glands which may be single (*Cadus*) or very many (*Zeus*). A rete mirabilia forms the vascular bases of the gland which is covered over the modified glandular epithelium invaginated into crypts.

In physostomi the secretion and absorption of gas is exceptionally slow and it is compensated by absorption and liberation of gas through the *ductus penumaticus*. The physoclistous fishes depend only on gas secretion and absorption through the walls of bladder. There is, therefore, a tendency on the part of the bladder to get differentiated in most of these fishes into an anterior oxygen producing and a posterior oxygen absorbing region. In the former special area, the red body or red gland for secreting gas is present and the later is thin to allow O_2, to pass to the capillaries over lying it. In specialized forms like *Mugil* the posterior region becomes converted into a flattened "oval" shaped arc which can be closed by a fold provided with a sphincter and dialator muscles.

The blood is supplied to the bladder in physostomi by a branch of the coeliaco-mesenteric artery and it is returned by hepatic portal vein. In physoclistous fishes the coeliaco-mesenteric artery and a vein from hepatic portal system constitute the blood supply to the red gland while branches from dorsal aorta and a branch from posterior cardinal supply the absorption region and anal region. In *Polyterus* and Dipnoi the blood is supplied to the bladder by arteries from the last pair or aortic arches and the blood is returned by hepatic veins in *Polyterus* and by pulmonary vein to heart in Dipnoi.

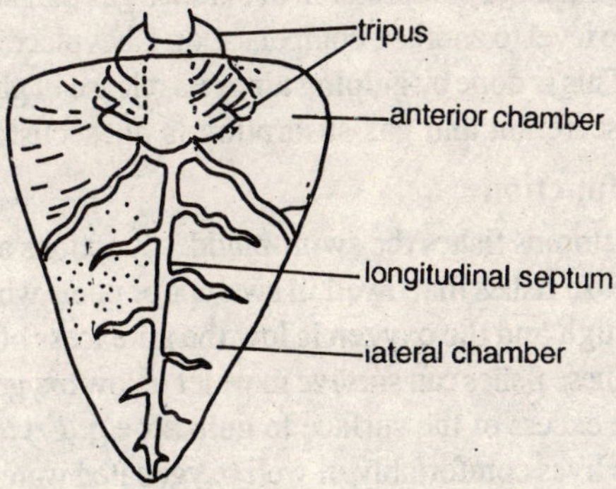

Fig. 9.3. Swim bladder of *Mystus*.

The swim bladder is innervated by sympathetic nervous system through a branch from coeliaco-mesenteric ganglion and by branches from the intestinal branches of vagus nerve.

A number of attempts have been made to derive the various kinds of air-bladders from some ancestral forms. *Boas* suggests that the original single dorsal bladder opened on the ventral side of oesophagus by two right and left ducts which reunited ventrally. According to *Sagenchi* the bladder was originally ventral and bilobed. It shifted round right side in Dipnoi and left side in Teleosts. *Erythrinus* in which the duct is on left side represents an intermediate stange in such a condition. According to *Goodrich* best solution is provided by *Spengel* the bladder or lung arose from the posterior pair of branchial pouches. The union of right and left pouches ventrally gave rise to condition of *Polyptreus*, Dipnoi and Tetrapoda while the more complete fusion dorsally yielded the air bladder of *Amia* and other Teleosts.

A variety of functions are associatd with the bladder:

1. Hydrostatic function

One of the principle function of swim bladder is that of a hydrostatic function by adjusting the amount of air inside it makes the density of fish more or less equal to that of its environment, so that it may swim with minimum effort. There are three lines of evidences in support of this conception, (a) Fish with well developed swim bladder swims easily and tends to remain poised in mid-water with great effort. (b) Density tension shows that there is a correlation between a well-developed air-bladder and low sinking factor and high sinking factor with the reduction or absence of swim-bladder. (c) The volume of swim bladder can be adjusted to compensate for expansion or compression of bladder-gas that occurs when fish moves from one level to another, compensation takes place by addition or removal of gas. This is done by gulping air or its release in physostomus fishes and by gas secretion and gas absorption is physoclistous fishes.

2. Respiratory function

In some physostomus fishes the swim-bladder functions as a lung or oxygen store-house. In fishes that dwell in swamps or pools where carbon dioxide tension is high and the oxygen is low the air excess of bladder is modified as lung. These fishes can survive in water of low oxygen contents provided they have excess of the surface to gulp air e.g. *Erythrinus* and *Umbra*. *Erythrinus* lives comfortably in well oxygenated water resulting to aquatic respiration but if the pool is poorly oxygenated and the fish is prevented from swallowing air the bladder is used as a lung.

The proportion of oxygen in air-bladder decreases when the fish is kept in water of low oxygen content. This has been demonstrated for a number of fishes like *Cyprinus*. It is assumed that the oxygen removed from bladder becomes available to the fish for respiration. The amount of oxygen according to *Hall* and *Khalin* removed in 6-8 hours would sufficient for 10-15 minutes. The extra oxygen available for respiration, therefore, not of much importance and the swim-bladder has a little significance as oxygen store. In deep sea the proportion of oxygen is higher and the actual amount of gas is greater as it is under pressure. In such a circumstances the swim-bladder may be important as oxygen store.

3. Sensory function

As compression and rarefaction of bladder gas occurs when the fish is subjected to pressure changes, the swim bladder put functions as barometer, hydrophone, manometer. If the bladder did such a function it could do either directly through stimulation of nerve ending lying in the wall of bladder or indirectly by its connection with the ear. It has been suggested that such a mechanism allows the fish to maintain a steady depth by responding to the changes of pressure by compensatory swimming movement.

Coupling of swim-bladder and ear is known from many species of fishes. The simplest method is by opposition of anterior part of bladder diverticulum on the fibrous membrane which covers the aperture of the outer wall of auditory capsule. Such a connection is known from fishes like *Notopterus*.

A more elaborate coupling occurs in Clupeid fishes. A pair of bladder diverticulae go to the head. Each diverticulum on entering the auditory capsule bifurcates and terminates into two sacs, anterior of which is enclosed in the bulla of periotic and post-carotid. The anterior sac comes in relation with the utriculus.

In certain other fishes a chain of ossicles called the *Weberian ossicle* link the anterior chamber of air bladder with the sacculus. The chain consists from behind forward are *tripus*, *intercalarium*. *scaphium* and *claustrum*. This connects swim-bladder, ossicles and sacculus together. A well developed area of skin against antero-lateral edge of bladder is without its musculature and is known as lateral cutaneous area. It is through this area passes impulse to bladder.

These ossicles increase fish sensitivity to volume. Changes of bladder allow finer control over escape of gas through the pneumatic duct. In other word these ossicles function as pressure resistor.

4. Sound production

Many fishes produce sound through the medium of swim-bladder may act as a resonator for sounds produced by the activity of neighbouring organ e.g. in *Balistis*. The bladder works as resonator for a sound produced by stridulation of the clythrum against postclaviely. The release of gas bubbles through pneumatic duct is often accompanied by sound as in *Cyprinus*. Sound production by contraction of special muscles which set-up vibration in the bladder wall is known e.g. *Auchenipterus* produces sound by a special muscle which runs from the occipital region of skull to the anterior of transverse process of 4th vertebra. A special muscle attached to skull by one end to other to bladder, produces sound in *Therapon*.

The sound produced appears to be characteristic for species and serves as reçognization mark which fascilates aggregation of breeding shoals. It is also assosiated with agressive behaviour. The species of *Osanus* gives grunt if disturbed with guarding nest. There is a possibility that deep water sounds or voices are used by fishes for echo sounding so that it may know at what depth it is swimming.

Swim-bladder and ecology of fishes

The principle function of air-bladder is hydrostatic and it is best developed in fishes which are active mid-water swimmer. The bladder is

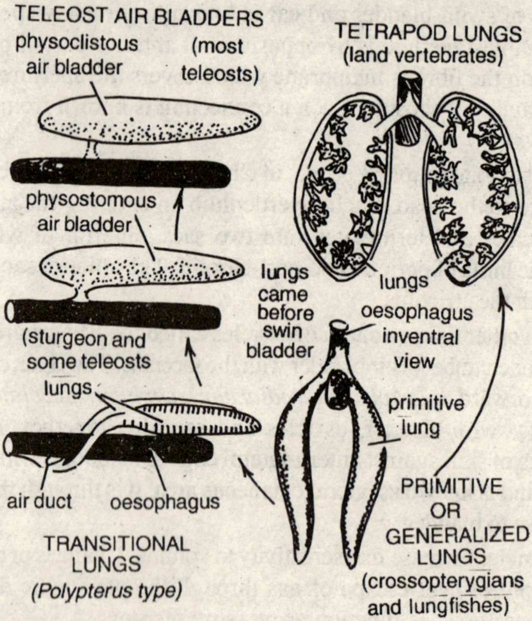

Fig. 9.4. Air bladder and lungs.

lost in fresh water species which are bottom living and may be reduced or lost in hill-stream fishes. In Ostariophysi the reduction affects the posterior chamber of bladder.

In neritic zone of the sea the swim bladder usually helps swimming between tide-mark. The swim-bladder present is closed in mid water dweller while in pelagic and coastal water it is open. The swim bladder is closed in surface dweller of oceanic zone and well developed in bathypelagic forms living upto 500 meters. The swims bladder is lost or converted into fat storing organ in deeper living species.

Physostomous predominates fresh water, there are a few of them in the neritic zone of sea and except for eels none in the oceanic water. Physostomous secretes gas very slowly and relight the pneumatic duct for the purpose. There being in deep water of sea would be unable to make long journey to the surface may would the lost if they did enough air is be taken to allow for compression in return to deep water. Closer of bladder and development of red gland have made these fishes independent of the surface but the closing of the bladder restriction is laid on the role of movement of fish from one depth to other. In very deep water so much to secrete gas into the bladder as to make it uneconomical. This may account for its lost or conversion as fat absorbing organ.

10

Accessory Respiratory Organs in Fishes

The principle organ of respiration in fishes is gills. Fishes are able to breathe oxygen dissolved in water by means of gills. The gills are richly supplied with blood capillaries. In some tropical fishes special structures are present for breathing in addition to gills. The structures are called the *accessory respiratory organs*. These structures are generally developed in response to the exceptional environmental conditions. The accessory respiratory organs rarely developed in the marine fishes and are reported by *George* (1953) in *Chiloscylium indicum* and *C. griseum* and *Achirus* (*Breader,* 1941). These organs are generally found in fishes living in fresh water viz. shallow stagnant water, pools, lakes, streams and rivers where deoxygenation of water occurs.

Some notable contribution on the accessory respiratory organs of fishes are those of *Das* (1927, 1949), *Breder* (1941), *Sawaya* (1946), *George* (1953), *Munshi* (1961, 62, 68), *Johansen* and *Hanson* (1968), *Johansen et.al.* (1970) and others.

About 140 species of fishes are reported to be adapted to air breathing among Teleosts. The following air breathing fishes are found in India.

S.No	*Family*	*Name*
1.	Notopteridae	*Notopterus chitala, N. notopterus.*
2.	Claridae	*Clarias batrachus*
3.	Cobitidae	*Lepidocephalichthys guntea.*
4.	Channidae	*Channa punctatus, C.gachua, C. striatus.*
5.	Heteropneustidae	*Heteropneustes fossilis*
6.	Amphipnoidae	*Monopterus cuchia*
7.	Anabantidae	*Anabas testudinius, Trichogaster fasciatus.*
8.	Gobidae	*Boleopthalmus boddaerti, Pseudapocryptis lanceolatus*
9.	Mastacembelidae	*Mastacembelus armatus, Macrognathus aculeatum.*
10.	Symbranchidae	*Symbranchus bengalensis.*

The following are the important accessory respiratory organs in fishes :

1. Wet skin

The wet skin of certain fishes acts as an accessory organ of respiration. It becomes richly vascularized and thin enough to facilitate exchange of gases. In the eel *Anguilla anguilla, Amphipnous cuchia* and in *Periophthalmus* and *Boleopthalmus.* The skin is thin when the fish is out of water. The gases exchange takes place through the wet skin as in the frog. They can pass through damp vegetation and cover a short distance. Thus making a land journey.

2. Buccopharyngeal epithelium

The mucous membrane of buccopharyngeal cavity serves as a respiratory organ. The membrane is highly vascular. It may be simple or may develop folds or pleats. It is seen in *Boleopthalmus, Periophthalmus, Electrophorus* and *Monopterus.*

3. Parts of alimentary canal

In certain fishes various parts of alimentary canal act as respiratory organs. The intestinal respiration is found in loaches (*Misgurnus fossilis*) and some cat fishes of Siuridae and Loricaridae. The *Misgurnus* swallows air and its intestinal wall extracts oxygen. They use intestine at the time of respiration. In Cobitidae, the middle and hinder parts of intestine serve for the both respiratory and digestive functions. In certain cases the intestine serves as the respiratory organs during the summer months only and during winters as digestive organ. The rectum of nesting catfish (*Collichthys*) absorbs oxygen drawn in through the anus and hindgut.

4. Saccular organs

In the cuchia eel, *Amphipnous cuchia* (Symbranchiformes) the gills are reduced and gill lamellae are present only on the second gill arch while the third arch bears a fleshy vascular membrane. The air breathing organs consist of a pair of sacs which grow out from the pharynx just above the gills. The air can be stored here for sometimes foı gaseous exchange. Each sac is provided with an inhalent and exhalent apertures. The respiratory epithelium of the sac consists of vascular areas in the form of islets bearing numerous *papillae* or *rosettes.* In *Hetropneustes fossilis* the saccular organ is also present. Here the sac is very long and reaches the tail region. The gills are fused to form a valve over the orifice of the sac. The islets are present and covered by a thin epithelium having numerous capillaries. The fourth afferent branchial artery supplies blood to the sac. and the fifth efferent branchial collects the blood.

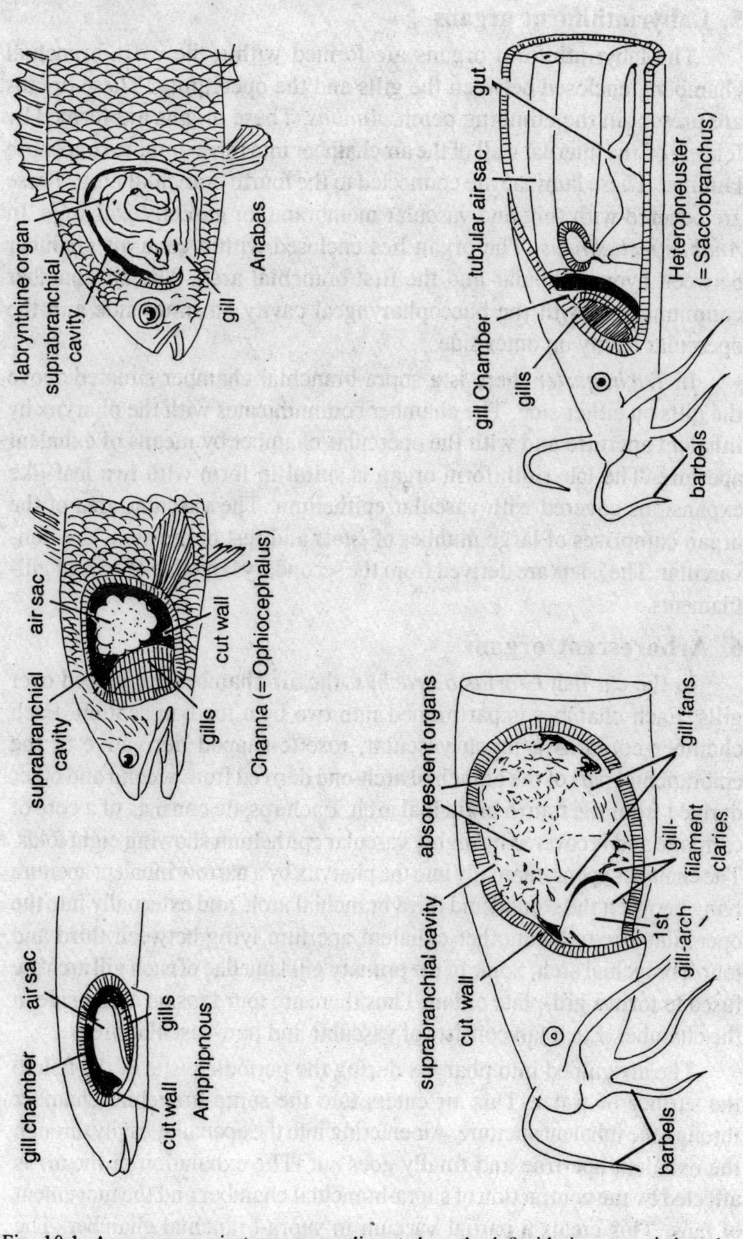

Fig. 10.1. Accessory respiratory organs dissected on the left side in some air-breathing teleost fishes.

5. Labyrinthiform organs

The labyrinthiform organs are formed within the extra-branchial chambers, enclosed between the gills and the operculum. These organs are present in the climbing perch, *Anabas.* These organs are formed by folding of the internal wall of the air chamber into three or more concentric laminae. These laminae are connected to the fourth branchial arch. These are covered with thin and vascular membrane for gaseous exchange. In *Anabas testudineus.* The organ lies enclosed within each air chamber between hyomandibular and the first branchial arch. The air chamber communicates with the buccopharyngeal cavity on inner side and the opercular cavity on outer side.

In *Trichogaster* there is a supra-branchial chamber situated above the gills on either side. The chamber communicates with the pharynx by inhalent aperture and with the opercular chamber by means of exhalent aperture. The labyrinthiform organ is spiral in form with two leaf-like expansions covered with vascular epithelium. The vascular area of the organ comprises of large number of *islets* and rest of the areas are non-vascular. The islets are derived from the secondary lamellae of typical gill-filaments.

6. Arborescent organs

In the cat fish *Clarias batrachus,* the air chamber are placed over gills. Each chamber is partitioned into two by a median septum. Each chamber contains a much vascular, rosette-shaped derivative of the epibranchial part of the branchial arch-one derived from second and other derived from the fourth branchial arch. Each rosstte consists of a core of cartilage and is covered by highly vascular epithelium showing eight folds. The chamber opens internally into the pharynx by a narrow inhalent aperture lying between the second and third branchial arch, and externally into the opercular cavity by another exhalent aperture lying between third and fourth branchial arch. Some of the primary gill lamellae of each gill arch are fused to form a gill-plate or fan. Thus there are four fans on either side in the chamber. Each fan consists of vascular and non-vascular areas.

The air gulped into pharynx during the periodic visits of the fish to the surface of water. This air enters into the supra-branchial chamber through the inhalent aperture. Air entering into the opercular cavity through the exhalent apertrue and finally goes out. The exhalation of the air is affected by the contraction of supra-branchial chamber and the movement of fans. This creats a partial vaccum in supra-branchial chamber. The mouth is opened and the buccopharyngeal cavity is enlarged to inhale air.

7. Accessory branchial chamber

The snake-heads (*Ophiocephalus striatus*), long, cylindrical fishes with slightly flattened and somewhat serpent like heads inhibit rivers and ponds as well as stagnant pools in the marshes. They have accessory branchial cavities on each side, above the gills. The cavity on each side is enclosed within a pouch which is derived as diverticulum from the pharyngeal wall. The cavity remains in communication with the pharynx by a narrow aperture. The inner lining of the accessory branchial chamber is highly vascular to permit exchange of gases. Other structures such as labyrinthine organs etc. are absent in this fish.

8. Air bladder

The air bladder of some fishes is modified for aerial respiration. Thus in *Amia, Lepidosteus, Polypterus, Gymnarchus* and in the Dipnoi, the air bladder is most highly developed as a lung. The air bladders are highly sacculates or alveolar in nature. In Dipnoi, the inner walls have become thin and richly vascular. It is partitioned by the development of septa and cross-septa to form a series of alveoli. The air bladder of *Notopterus* has a wide pneumatic duct and acts as accessory respiratory organ. The main part of air bladder lies in the abdominal cavity and bifurcates into two lateral bronchi-like tubes. In the trunk region from the bladder 14-15 blind pouches with finger-like processes are given out. These pouches and their processes are richely supplied with blood capillaries.

11

Sound Production organs & Weberian Ossicles

There is a widespread and popular belief that fishes do not produce sounds. This is quite untrue, although incapable of vocal efforts comparable to those of mammals and birds, a number of bony fishes produce sounds of one sort or another, and some forms are provided with special sound producing organs. These organs may be associated with the swim bladder, fin spines, vertebrae, and so on, and provide another example of the assumption of new duties by organs originally employed for a totally different purpose, The fishes produce sounds in various ways, by grinding their teeth, by rasping spines and fin-rays, or by burping, faiting or gulping air.

The sound production may be discussed under three heads:

1. Stridulating Sounds

Stridulation as a method of sound production is known from many fishes. It is made possible simply by rubbing one surface against another. For example. the Hase Mackeral (*Trachurus*), the Sun fish (*Mola*), and certain species of Trigger fishes (*Balistes*) produce harsh noises by grating together the upper and the lower pharyngeal teeth. The Bull head (*Cottus*) uses a portion of the gill cover for stridulation; the Flying gumard (*Dactylopterus*) uses the hyomandibula bone; the Trigger fish (*Balistes*), File fish (*Monacanthums*), Boar fish (*Capros*), Sturgeon fish (*Acanthurus*), Stickle back (*Gasterosteus*) use spines of the dorsal, anal, pectoral or pelvic fins for sound production.

In *Rhmecanthys aculeatus* the noise is said to be due to the friction of certain of the bones of the arch supporting the pectoral fin against one another, and since these are more or less intimately associated with the swim bladder, the latter acts as amplifier and intensifies the sound vibration.

2. Breathing Sounds

The breathing sounds are produced by few bony fishes like Cyprinus and Siluras. The sound vibrations are said to originate by the expulsion of gas from the bladder through the ductus pneumaticus and mouth. In *Misgurnus fossilis* sound is produced by rapid explusion of air through anus.

3. By Muscles Connected to Air Bladder

In many cases the noise is produced through the agency of special muscles associated with the swim bladder. In a number of cat fishes (Siluroidei) an apparatus known as the elastic spring mechanism occurs, the purpose of which is to cause the walls of the swim bladder to vibrate. The springs are specially modified portion of the fourth vertebrae and their expanded ends are attached to the front part of the swim bladder. Two strong muscles run from the springs to the hinder portion of the skull, and when these contract and springs, and with them the walls of the bladder vibrate rapidly and produce sort of growling or humming noise. Generally the swim bladder is divided up by internal prtitions into a number of chambers and freely communicating with one another, and there can be no doubt that the sound is intensified by the vibrating movements of the gases contained in the bladder across the free edges of the partition.

In drum fishes or *Sciaenids* renowned for their vocal efforts, the sounds are produced by the rapid vibration of special muscles, which are not always attached directly to the bladder, but may run from the abdomen on either side to a central tendon situated above the bladder. The rapid contraction and expansion of the muscle causes the walls of the bladder to vibrate, and since this has a complicated structure, it acts as a sort of resonator and intensifies the sound.

Certain deep sea Rat tail (Macrouridae) have swim bladders whose structure suggests that the fishes are capable of sound production. The actual noises produced by the different fishes present great diversity, ranging from a more or less melodious vocal effort to a mere grunt. In the Gurnards (Triglidae) the sounds are of a somewhat different nature, and have been variously described as grunting, crooning, snoring etc. Unlike the Drums, these fishes do not make use of rapidly repeated sounds and rolls, but produce short, sharp sounds repeated at more or less lengthy intervals. The grunt can be imitated, by drawing the fore fingers and thumb towards each other over the surface of an inflated rubber balloon.

It has been concluded that sound produced by fishes serves for communication especially during the breeding season and may also be used for offense. A toad fish, probably has a mating call. It is also suggested that, particularly for fishes living over deep water, that the sounds may function rather as an echo-locater and thus help to fix the position of fish in mid water.

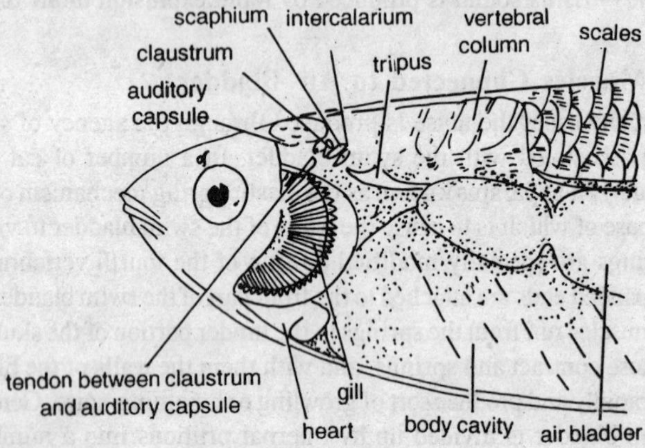

Fig. 11.1. Weberian ossicles and air bladder.

WEBERIAN OSSICLES

In the Ostariophysi (Cypriniforms) a chain of small ossicles of vertebral origin link the swim bladder with the sacculus at the base of the ear. These ossicles were described and their function understood by *Weber* as long ago in 1820, and they are rightly still termed the *Weberian apparatus*, though some of the names he gave to the seperate ossicles have now been changed. *Weber* considered the ossicles of mammals and, therefore, he named them as *claustrum*, *stapes*, *incus* and the *malleus*. In the year 1889, *Bridge* and *Haddon* found a considerable variations in the development, form and functions of these ossicles and considered them to be organs that help in registering hydrostatic changes. Thenafter their homology with the mammalian ossicles was disproved and designated as *claustrum*, *scaphium*, *intercalarium* and the *tripus*.

Structure

As stated already that Weberian apparatus is composed of a chain of four ossicles-claustrum, scaphium, intercalarium and the tripus. The *claustrum* is embedded with side in the inter-ossicular ligament and articulates with a part of the neural arch of Ist vertebra. Its lower end lies on the inner concave side of scaphium. The *scaphium* is slightly bigger and broader than the claustrum. It is a compressed structure. A ventral

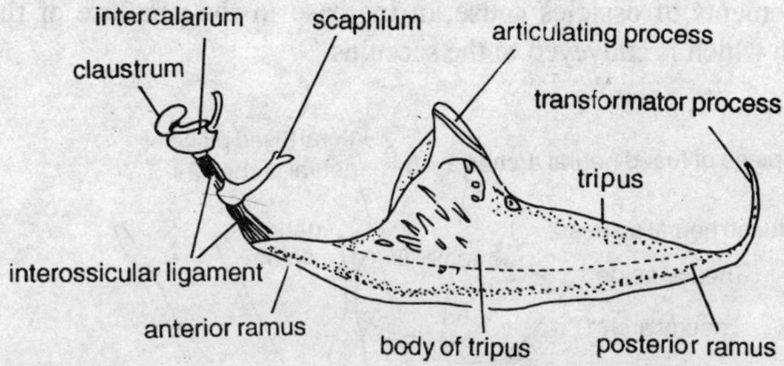

Fig. 11.2. Chain of Weberian ossicles.

peg-like process of the scaphium may be present and articulate with a depression in the centrum of the first vertebra. To connect the scaphium from its inner side there are inter-ossicular ligaments. *Intercalarium* is reduced to a nodule of one which lies in the inter-ossicular ligament, covering its full width. Intercalarium is not connected with vertebral column as in the case of *Wallago* and *Mystus*, it may be in the form of a rod-like piece reaching upto the centrum of the second vertebra, as in the *Labeo*, *Cirrhina* and *Tor*. The end of intercalarium, not connected with ligaments, is bifurcated. To connect the intercalarium and tripus together the ligaments are developed there. The *tripus* is the largest Weberian ossicle and consists three process. Anterior process articulates the inter-ossicular ligaments, the middle one articulates between the second and third vertebrae. The posterior process is slightly curved and is also called the transformator and is connected with the wall of the gas bladder. The right and left sacculi of the membranous labyrinth are connected by means of a fluid filled transverse canal. Form the middle of this canal, a blind diverticulum the *sinus endolymphaticus*, projects posteriorly into a central fluid filled cavity. A small diverticulum of this cavity the *atrium sinus impairs*, containing perilymph, lies between the claustrum and the scaphium of the Weberian ossicles.

Functions of the Weberian Ossicles

In the siluroids there is a direct mechanical transmission of the vibrations. In these the swim bladder is enclosed in a bony capsule and protrudes through an opening in it for attachment with the tripus. Any change in the volume of the bladder causes its walls to bulge out of the opening and forces the ossicles forwards. In cyprinoides there is an indirect transmission of vibrations. This method involves a backward movement of the posterior process of the tripus resulting its anterior movements. The movements of ossicles cause an increase in the pressure of the perilymph which is conveyed to the sacculus.

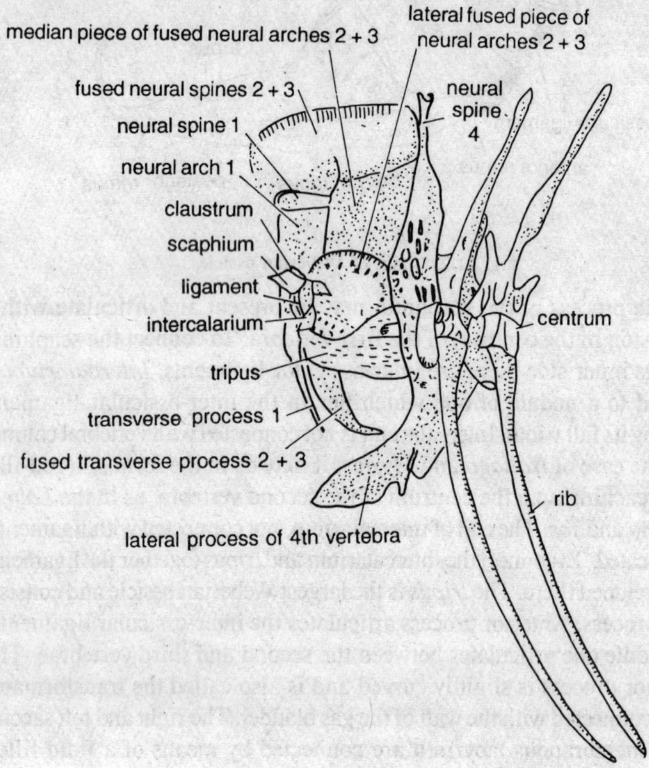

Fig. 11.3. Weberian ossicles attached to vertebrae.

According to *Bridge* and *Haddon* the ossicles are sensitive to changes in the volume of the swim bladder due to variations in the hydrostatic pressure. Any change in the volume of swim bladder causes movements of the ossicles, which transmit the change in the pressure to the sinus impart. This is then conveyed to the sacculus. This is the hydrostatic function of the Weberian ossicles.

The Weberian ossicles also help in detecting the variations in the atmospheric pressure. Thus it works as a barometer. Some species like the loaches are very sensitive to changes in the atmospheric pressure.

According to *von Frisch* and others the Weberian ossicles also perform an auditory function. The ossicles transmit some vibration to the sacculus.

It has been observed that the fish whose ears are connected to the swim bladder are sensitive to sounds of higher frequencies and lower intensities than those where this connection is not developed.

According to *Evans* the Weberian ossicles also help in the echolocation and in the detection of the direction of coming sound vibrations.

12

Electric Organs in Fishes

No animal except fishes are known to possess organs with special function to produce electricity. It is the first bio-chemical potential observed by man. In certain fishes alone, in Animal Kingdom, the energy produced by vital processes assume the electrical form. About 250 species of fishes are reported to possess electrical organs. The sharks were known to give an electric shock since ancient times. Towards the end of 18th century *Anderson* suggested the electrical discharge by sharks from leden jar. *Faraday* made crucial test with galvanometer and spark gas on electrophorus, thus demonstrating the electrical nature of the sharks. The list of electric fishes was further extended in the 19th century. Some important fishes are being listed below.

Table : 12.1.

	Group	*Genus*	*Distribution*
I.	Teleosts		
	Mormyridae	1. Gymnarchus	From W. of Africa
		2. *Mormyrus*	From W. of Tropical Africa
	Gymnotidae	1. *Electrophorus*	From W. of South America
		2. *Gyminotus*	Central and South America
	Siluridae	1. *Malapterurus*	From W. of Tropical Africa
		2. *Astroscopus*	Atlantic and Pacific coasts of Southern North America
II.	Elasmobranch	1. Electric Rays	All oceans
		2. Skates	All temperate oceans

STRUCTURE OF ELECTRIC ORGANS

The position of electric organs in electric fishes varies from species to species. They resemble structrally in all case. Each organs is composed of large number of hexagonal plates called *electroplates* or *electroplaxes*.

These plates are more or less orderly arranged in such a way that all face in one direction and thus give additive effect. Each electroplate is embedded in a jelly like material and enclosed in a compartment of connective tissue. The electroplates are innervated by the nerve fibres while the jelly receives blood capillaries. The electroplates are unicellular and each cell is a multinucleated with nearly transparent cytoplasm. The electric organs are clean gelatinous masses which are readily be distinguished from surrounding muscles. As a rule the surface on which the nerve terminates is smooth as compared to the other surface, that bears papillae. This is probably an adaption to imporve the efficiency of electric discharge by lowering the effective internal resistance of the electroplate due to increase in the total membrane area. The mode of contact between nerves and electroplates varies from species to species. In *Torpedo* and *Raja,* the nerves branch repeatedly and extend almost over the whole surface of the electroplate. In *Electrophorus* the nerves end on a number of short papillae there by reducing the area of contact. In *Mormyrus* the area of contact is further reduced and in *Malapterurus,* the contact at one point only has been observed.

Electric organs in *Torpedo*

The electric organs in *Torpedo* are present in the form of two large and two small organs on each side of the head. The large one encloses the smaller one. The organ is composed of hexagonal columns which are as many as 500. Each column has got about 476 plates. Each column is negative on its ventral side and positive on dorsal side. The current passes from the dorsal side to the ventral side i.e. from positive to negative. The

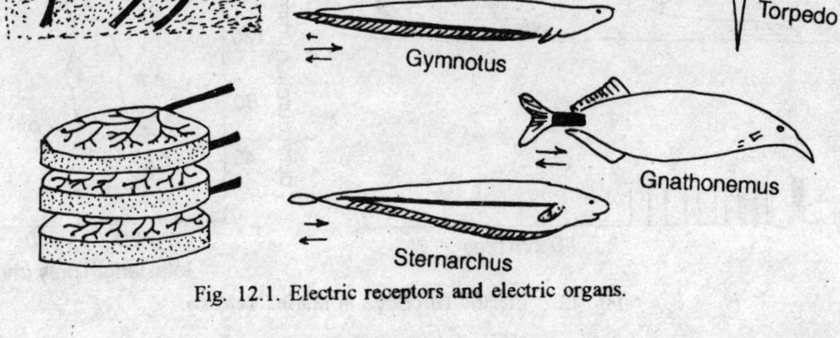

Fig. 12.1. Electric receptors and electric organs.

discharge for *Torpedo* is about 30 volts, sometimes reaching upto 50 volts. The fish may release as many as 150 discharges in a second. The shocks are sufficently poweful to kill other fishes upto 4 to 5 lb. in weight.

Electric organs in *Electrophorus*

The *Electrophorus* is a common electric eel. It is as long as 8 feet and is capable of giving powerful shocks. Its electric organs are huge and are on either side, each organ consists of column of cells arranged longitudinally occupying nearly half the mass of the body. There are about 70 compartments each one contains about 6000 electroplates. Spinal nerves innervate the electric organs. The front end of the column being the positive and the other end negative. The discharge occurs at a rate of several hundreds a second producing a current of about 500 volts. This current splits up body water into hydrogen and oxygen. The oxygen is utilized in breathing and hydrogen is removed out in the form of bubbles through the mouth. The current flows from the tail to the head. Its shocks are quite severe and can kill small animals and paralyse big animals like cattles.

Electric organ in *Malapterurus*

Malapterurus is a common African electric catfish. Electric organs are located in the skin where they form a thick but semitransparent jacket loosely covers the entire length of the fish. The organs are divided into compartments by septa of connective tissue. These contain electroplates whose nerves ends are towards the tail. The polarity in this fish is from tail to head. The plates are innervated by the spinal cord.

ORIGIN AND DEVELOPMENT OF ELECTRIC ORGANS

In all fishes except *Malapterurus* it has long been known that electric organs are modified muscle fibres. In *Torpedo,* the entire electric organ

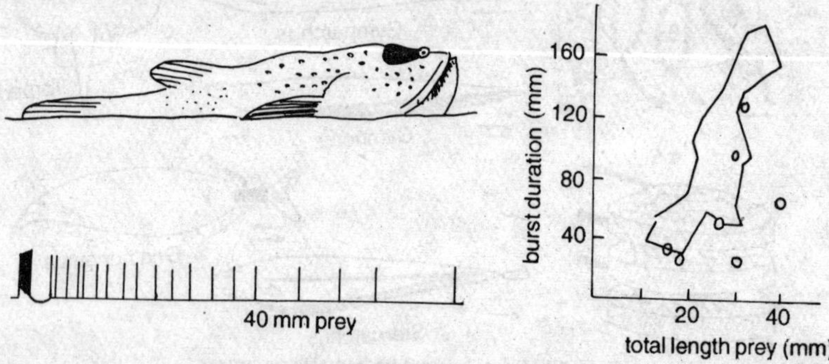

Fig. 12.2. Electric Discharge in Marine Teleosts.

has been derived from some of the branchial muscles which have lost their original function of moving the gill-arches. The organ of electric eel, the skates, and Mormyridae are similarly modified muscles and owe their origin similar to the transformation of some of the lateral muscles of the tail. In star-gazer, the electric organs have been shown to be developed from portion of the eye muscles and each of the plate represents a single muscle fibre.

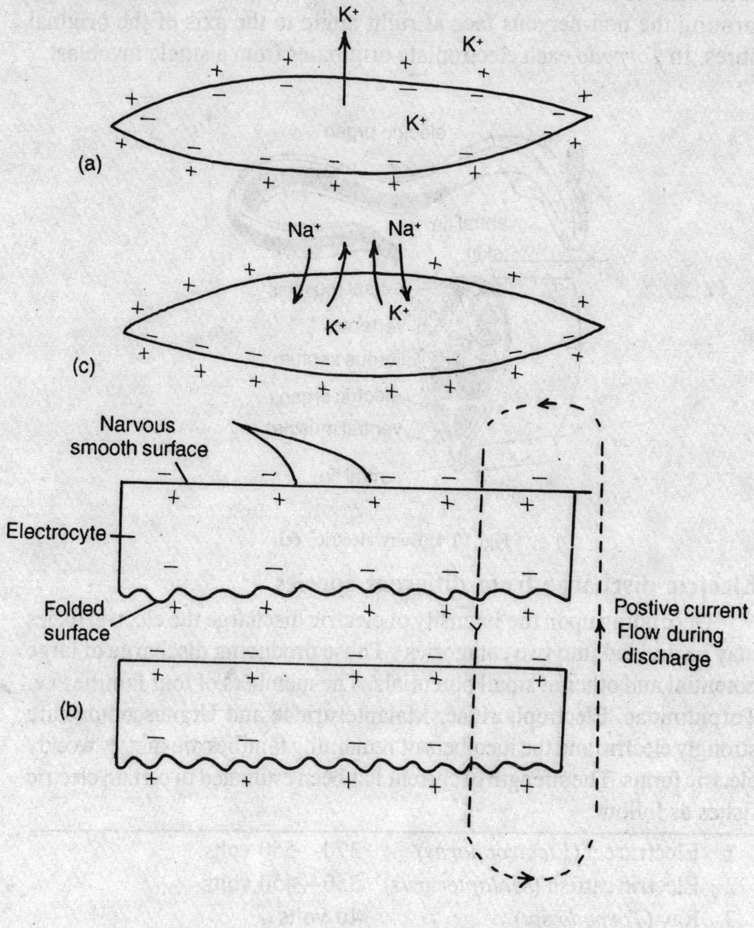

Fig. 12.3. Showing mechanism of producing discharge. (a) showing resting potential; (b) when the electrocyte is stimulated (c) showing Na^+ influx across the nervous surface K^+ leakage across the rough surface leading to positive current flow to the direction indicated.

Development

The best known studies of the development of electroplates are those of *Ewart* (1892) for *Rais batia.* In the youngest embryo the electric organ was filled with muscle fibres not visibly different from those around them. In 7mm embryos the anterior end of the fibre began swelling and in 11mm embryos the fibres had become club shaped. The innevation was now concentrated at the upper end of the nervous face of the electroplate. In later embryos the muscle tail on the posterior side gradually divided away forming the non-nervous face at right angle to the axis of the original fibres. In *Torpedo* each electroplate originates from a single myoblast.

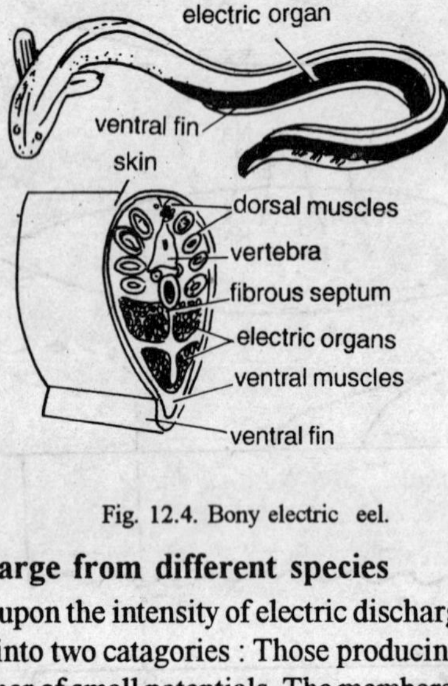

Fig. 12.4. Bony electric eel.

Electric discharge from different species

Depending upon the intensity of electric discharge the electric fishes may be divided into two catagories : Those producing discharge of large potential and other of small potentials. The members of four families i.e. Torpidinidae, Electrophoridae, Malapteruridae and Uranoscopidae are strongly electric and the members of remaining families are mostly weekly electric forms. The strength of current has been estimated in certain electric fishes as follow :

1. Electric eel (*Electrophorus*)	370—550 volts
2. Electric catfish (*Malapterurus*)	350—450 volts.
3. Ray (*Torpedo sp.*)	40 volts
4. Ray (*Narcine sp.*)	37 volts
5. Star-gazer (*Astroscopus*)	50 volts
6. Skate (*Raja clavata*)	4 volts

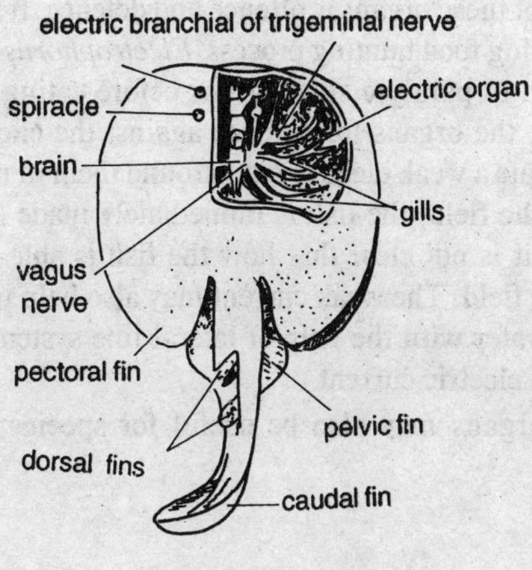

Fig. 12.5. *Torpedo.*

Mechanism

The exact mechanism and function of electric organs in producing and storing the electric current is not fully understood. The current is generated by the modified muscle end plate on the nervous side of electrocytes. At rest the two nervous and alveolar surfaces of electrocytes remain positively charged on outer side and negatively charged inside. It is because, the permeability of membrane to potassium ions (K^+) exceeds the sodium ions (Na^+). It results in faster exit of K^+ outwards than the inward diffusion of Na^+. At this stage the resting potential is about 100mV. As soon as the stimulus reaches to the nervous face of electrocyte, the face becomes permeable to Na^+ as a result Na^+ enter the electrocyte. This depolarizes the face now it becomes negatively charged on outer side and postively charged on the inner side. The potential at the alveolar end remians the same. On inner side the charge measures about 60mV. As a result there is a total potential difference of 160 mV between the two surfaces. The voltage of successive electroplates are added up in series and a strong current flows.

The polarisation of electric organs and the direction of current flow is not the same in all the electric fishes. In *Torpedo* the current flows from drosal to ventral side. In *Electrophorus*, the current flows from tail to head region.

Functions of Electric organs

The difference in position, structure and strength of the electric organs in various species are probably due to difference in their functions. The principle function of these organs is offence and defence. It is used as the offensive organ during food hunting process. *Electrophorus* and *Torpedo* have been observed to paralyse small fishes before eating them. Some fishes *Narcine*, use the organs for defence against the enemies. Fishes like mormyrids create a weak electric field around them in muddy water, any object breaks the field, the fish is immediately made aware of this change. However, it is not clear that how the fish is able to detect the break in its electric field. The weak current may also help in finding the direction in dark water with the help of lateral line system as the later help in detection of electric current.

The electric organs may also be useful for species or even sex recognition.

13

Endocrine Glands in Fishes

Fishes possess almost all the endocrine glands which are found in other vertebrates, but their histophysiological aspect is not well known. More important glands are pituitary, thyroid, adrenals or supra-renals, pancreatic islets, thymus, ovary, testis and the parathyroid or ultimobranchial body. A few like pineal glands, corpuscles of Stannius, urophysis and intestinal mucoss are considered to be doubtful nature.

PITUITARY GLANDS (HYPOPHYSIS)

The study on pituitany gland and its hormones in fishes have been extensively made by *Gorbman* (1939), *Olivereau* (1954, 1963, 1967), *Ramaswami* and *Sundararaj* (1957), *Porte* (1961), *Wingstrand* (1966), *Ball* (966), *Ibraham* and *Chaudhuri* (1970) and others.

The pituitary is situated on ventral side of the brain just behind the optic chiasma and is attached to the infundibulum by a stalk. In some cases the stalk is absent and the gland is pressed closely against the floor of the brain while in *Lophius* it is extremely long. In all vertebrates, the pituitary gland cosnist of two parts, separable on the basis of embryology, structure and function. These are *adenohypophysis* originating as an ectodermal upgrowth (Rathke's pouch) from the roof of the embryonic buccel cavity and the *neurohypophysis,* a downgrowth from the floor of the diencephalon. The two parts meet and enclose between them a mesodermal rudiment which give rise to their intrinsic blood vesseles. Thus the pituitary gland is a composite organ, and it has many different endocrime functions.

The adenohypophysis is the site of synthesis, storage and release into the blood several different peptide hormones; and the grater part of pituitary histophysiology is concerned with the allocation of each of these hormones to the type of pituitary cells that secretes it. The adenohypophysis is divided into two parts *pars distalis* and *pars intermedia.* Pars distalis in further divided into *rostral pars distalis* (*Proadenohypophysis*) and *proximal pars distalis (mesoadenohypophysis)*

Rostral pars distalis is the anterior most region of adenohypophysis and is populated mainly by the *acidophil cells.* These cells are stained by azocarmine or orange-G. Besides few *cyanophil cells* are also present which are stained by aniline blue and aldehyde fuchsion (AF). Few *chromophobes* are also present.

The proximal pars distalis is situated between rostral pass distalis and pars intermedia. This region shows cylic activity ie. increases or decreases in size during different periods of the year. It contains *acidophils, cyanophils* and *chromophobes.* The acidophil cells are more numerous during rest of the year while the cyanophils are most active during maturation and spawning periods.

Pars intermedia occurpies the lower part of pituitary gland. It partly or completely surrounds the distal part of neurohypophysis. It varies greatly in size in different species. Two types of cells are seen *cyanophils* stained by aniline blue and *acidophils,* stained by orange-G. In sone species amphiphils are also present taking both the stains. Besides above cells, there may the few *chromophobe* cells.

The neurohypophysis in fishes is rather simplar than the terrestrial vertebrates. It consists essentially of a hypophysial stalk, suspending the gland from the ventral region of diencephalon and containing an extension of the third ventricle (infundibular recess) and at the distal end of the stalk an enlargement, the *neurohypophysial lobe*, forms the middle of the gland. The stalk contains the axonal fibres of neurosecretory cells and the cytons or cell-bodies being located in the hypothalamus. The neurohypophysial lobe consists largely of the endings of these fibres interspread with cells called pituicytes. The neurohypophyis is a storage-release centre for the materials which are actually synthesized in the hypothalamus and than transpoted to the neurohypophysial lobe along the neurosecretory axons.

Histophysiology of Pituitany gland

In the adenohypophysis, there are several types of cells that take different stains. These are as follows :

Erythrosinophilic cells or Paralactin cells

1. ***Prolactin cells.*** These are also called the η cells. The cells are present rostral pars distalis. The cells are arranged in follicles. The cells are columnar in appearance. Their cytoplasmic granules take red stain with azocarmine or acid fuchsin. They are negative to paraldehyde fuchsin

(PAF) and alcine blue stains. These cells secrete a hormone called *prolactin*. In fishes there is no question of milk production, this hormone has got other functions. It is probably concerned with osmoregulation. Without pituitany, some teleosts can not survive in fresh water, though they can continue for indefinite period in sea water (*Betta splendens* and *Tilapia mossambica)*. Prolactin thus may be concerned with electrolyte balance. According to *Pickford* and *Kasto* (1957), this hormone is also concerned with melanogenesis.

2. *ACTH cells (Corticotrops or X-cells or ε - cells).* These cells are present in a sheet at the interface between the rostral pars distalis and neurohypophysis. The cells are small and the cytoplasm is nongranular, however, in some species of eells the cytoplasm is granular. The cells are usually *chromophobic* and can be stained by erythrosin or azocarmine or alizarin blue. They are elongated cells, usually arranged with long axis perpendicular to the adeno-neurohypophysial interface. The nucleus is elnogated with a distinctive scattering of chromatin.

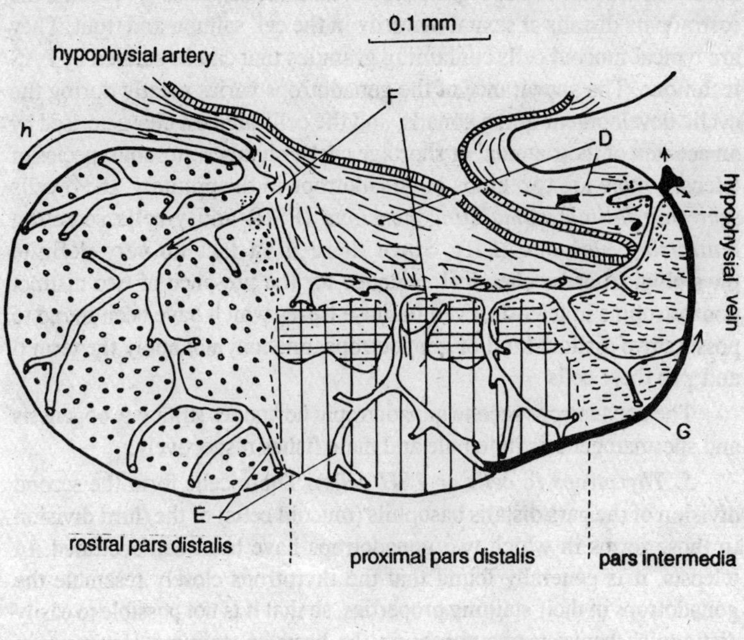

Fig. 13.1. Diagram to illustrate the Main Features and the Blood Supply to the Cyprinodont Pituitary. Course of blood circulation through, D→E→F→G→ hypophysial vein.

These cells secrete adrenocorticotropic hormone (ACTH). The hormone stimulates adrenocortical cells to secrete mineralocorticoids and glucocorticoid hormones like that of mammals

3. Growth hormone cells (Somatotrops or α- cells or GH cells or STH cells). These cells are the most prominant acidophils in pars distalis, especially in marine fishes with inactive prolactin cells, and occupy much of the proximal pars distalis. They can be easily stained with orange G. In *Poecilia* the η-α tinctorial distinction can be made quite easily with azan or alizarin blue method the η cells staining red with azocarmine and α cells stainig orange. In *Anguilla* the somatotrops are intermingled with the thyrotrops.

These cells secrete *growth hormone* (GH). It regulates the growth of various organs. It increases protein synthesis, decreases nitrogen loss and stimulates oxidation.

4. Gonadotrops (β and γ cells). The gonadotrops of Teleosts have usually been found located in the proximal pars distalis, and more often in the ventral part of this region. However, these cells also spread into the rostral pars distalis at sexual maturity in the eel, salmon and trout. They are typical mucoid cells containing granules that can be stained by PAS technique. The appearance of the gonadotrops varies greatly during the cyclic development of the gonads, and the cells are best characterised by an account of their annual or shortage cyclic changes. In some species of teleosts, there are two kinds of gonadotrops in the pituitary, the β-cells secreting *follicle stimulating hormone* (FSH) and γ-cells secreting *leutinizing hormone* (LH). Since there is in fact no very definite physiological or biochemical evidence for the presence of two distinct gonatotropins in teleosts, even in those fihses which have been found to possess two distinct kind of gonadotrops, one may not apply the term β and γ to these cells.

The gonadotrops secrete gonadotropic hormones stimulate oogenesis and spermatogenesis in female and male fishes respectively.

5. Thyrotrops (δ cells or TSH cells). These cells form the second division of the pars distalis basophils (mucoid cells) or the third division in those forms in which two gonadotrops have been demonstrated. In teleosts, it is generally found that the thyrotrops closely resemble the gonadotrops in their staining properties, so that it is not possible to easily distinguish between two purely on the basis of staining. However, in minnow (*Phoxinus)* the gonadotrops are AF negative but the thyrotrops are AF positive. (*Barrington* and *Matty*, 1955). These two types of cells also differ in their position also. The thyrotrops lie dorsally in the proximal pars distalis, while the gonadotrops lie in ventral region of some species.

Thyrotrops secrete thyroid stimulating hormone (TSH) which stimulates the thyroid gland and controls the production of thyroxine hormone.

Pars intermedia

It partly or completely surrounds the distal and largest region of neurohypophysial lobe and is usually extensively invaded by neurohypophysial tracts. Two cell types may usually be distingnished by their staining reactions and also by differences in shape and position. These are *cyanophils* and *acidophils*. Besides *amphiphils* and *chromophobe* cells may also present in few species.

Presence of two distinct types of cells suggests the production of two distinct hormones by pars intermedia. These are *melanophore stimulating hormone* (MSH) causing melanophore dispersion and the other hormone is *melanophore concentrating hormone* (MCH) inducing melanin concentration in the middle of pigment cells. This two *hormone hypothesis* is not universally accepted. The lack of certainty abont whether the pars intermedia produces one or two melanotropic hormones is reflected by uncertainty abont the functions of the types of cells in the pars intermedia.

Neurohypophysis

The neurohypophysis is composed of nerve fibres, mostly nonmyelinated, orginating from the neuronal cells bodies in the hypothalamus. These nerve fibres extend down the pituitary stalk into the pituitary gland to form the main bulk of neurohypophysis. Many fibres terminate in the neurohyprophysial lobe in close relationship to the blood vessels and many other fibres extend beyong the lobe and invade all regions of the adenohypophysis scattered ependymal cells and glial cells occur thronghout the neurohypophysis.

A special feature of fish pituitary is the extensive penetration of the pars intermedia by neurohypophysis, the two regions mixing in complex interdigitions. Two types of neurohypophysial fibres may be seen, depending upon the material contained in fibre will stain with classic neurosecretory stains such as AF, chrome alum hematoxylin, aldehyde thionin and AB. The stainable fibres are named *Type A* and nonstainable as *Type B*. Type A- fibres probably originate from the nucleus preopticus (NPO) of hypohalamus and the Type B fibres from nucleus lateralis tuberis (NLT) of hypothatamus. Neurosecretory material in both the fibres may accumulate in neurohypophysis in large masses termed *Herring bodies.*

Evidences show that at least seven neurohypophysial peptides have been identified in teleosts. The first group of these peptides is associated with antidiuretic vasopressor principles and concerned with water balance including control of urine and ionic regulation, The second group of these peptides is associated with oxytocian like principles and is respobsible for the contraction of oviducts and uterine muscles.

THYROID GLAND

The thyroid gland is normally present in close association with the ventral aorta and branchial arteries in the teleosts : Sometimes unusual position have been reported. There is great variations in shape and anatomical position. In *Myxine* and in a large majority of the teleosts, the thyroid follicles are not assambled into a single gland, but instead they are scattered generally in connective tissue throughout the sub-pharyngeal area. The thyroid follicles can disperse from pharyngeal region to eye, brain, heart, spleen, kidney and other tissue of the body (*Bakekr-Cohen*, 1959).

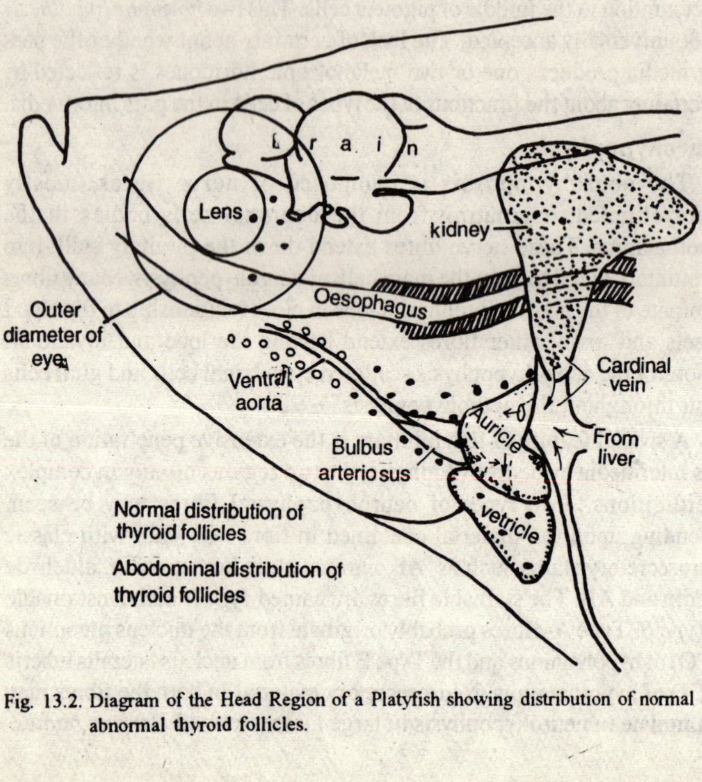

Fig. 13.2. Diagram of the Head Region of a Platyfish showing distribution of normal and abnormal thyroid follicles.

These heterotrophic follicles have been considered to be functional. *Chavin* (1956) confirmed that the head kidney thyroid is more active than pharyngeal thyroid. Several indian air breathing fishes as *Channa punctatus, C. gachua, C. striatus* and *Heteropneustes fossilis* also possess a compact and encapsulated thypoid close to the ventral aorta.

Histologically, the thyroid gland composed of a large number of follicles. Each follicle comprises a single layered epithelium enclosing a fluid filled space. They are bound together by connective tissue containing a large number of blood capillaries. The epithelium of follicle varies from low to high cuboidal in hight, depending upon their secretory activity. The lumen of each follicle is full of colloid which may be basophilic or acidophilic depending upon the secretory activity of the follicle and may show vacuoles in it.

The hormone secreted by thyroid gland is *thyroxine*. The thyroxine is an iodinated derivative of amino acid tyrosine. The thyroid follicle extracts iodide from blood, and its active secretion through follicular cells into the lumen of follicle. In the lumen, the iodine is formed from iodide by its oxidation. This iodine is bounded to a protein called *thyroglobulin*. The throglobulin is splitted by a protease enzyme leads to the formation of four different kinds of iodine bearing molecules. The *mono-iodotyrosine* and *diodo-tyrosine* contain one molecule of tyrosine and one and two iodine molecules respectively. Form these *tri-iodotyrosine* and *tetra-iodotyrosine* or *thyroxine* are derived by further condensation. The mono- and di-iodotyrosine are deiodinated within the follicular lumen while the tri-and tetra-iodotyrosine are released in the blood circulation. Thyroid hormone is responsible for several metabolic activities in fishes including growth and development. It is responsible for protein and carbohydrate metabolism, calorigenesis and calcium metabolism. According to *Muller* (1953) thyroxine also plays important role in oxygen consumption, but some contradictory reports have been brought into light regarding oxygen consumption. Thyroxine has been considered important during fish migration. The thyroid gland seems to be active during migration period of fishes. Scales and bone formation, the electrolytic movements are also known to be influenced by the thyroxine.

ADRENAL GLANDS

The adrenal glands of the most vertebrates and mammals consists of outer *cortex* and inner *medulla*. The cortex represented by interrenal cells is derived from the mesoderm and produces steroidal hormones while medulla represented by chromaffin cells is derived from the ectoderm of neural crest and produces amines.

Interrenal tissue

In cyclostomes the interrend cells are scattered along the wall of posterior cardinal veins and in the mesonephros. The cells are present in several layers. In *Puntius* these cells form thick glandular mass while in *Channa punctatus* they are arranged in the form of lobules along the posterior cardinal vein, In *Heteropneusteus fossilis* the cells are arranged in lobules varying in thickness and its branches in the hemopoeitic head kidney. The interrenal cells are generally columnar or polygonal. The cytoplasm is acidophilic with a distinct nucleus. During recent years the presence of steroid converting enzymes, steroid precursors and corticosteroids have been shown in the interrenal cells. The internenal gland of some teleosts can convert pregnenolone and progesterone into cortisol and cortisone.

Chromaffin tissue

Distribution of the chromaffin tissue in fishes show wide variations. They may be associated with the interrenal tissue. The chromaffin cells are found scattered in the head kidney of several teleosts. They may occur singly or in groups among the interrenal cells or associated with interrenal tissue to form an complex comparable to the adrenal gland of higher vertebrates. The cells are large and irregular containing a central nucleus. Chromaffin material is present in the cytoplasm closely surrounding the nucleus.

The hormones of inter-renal tissue are collectively known a *adrenal corticalsteroids*. The most potent corticoid is the *aldosterone*. This hormone regulates the sodium level by acting on gills and on kidney tubules. Other corticoids like *cortisone*, *corticosterone* and *cortisol* have also been found. These influence proteins, fats and carbohydrates metabolism. Hypoglycemia and lowering of liver glycogen have been demonstrated in eels when the interrenal gland is removed. The interrenal with increased secretion of cortisol influences the maturation of oocytes in *Heteropneustes fossilis*.

The chromaffin tissue secretes *epinephrine* and *norepinephrine*. When the fishes are disturbed for sometimes, there is abnormal raise of epinephrine in the blood plasma, Epinephrine increases the plasma fatty acids of eels, sea-bass and scorpion fish but in gold fish, carp and beam - the plasma fatty acid concentration decreases. In trouts epinephrine increases cyclic- AMP concentration.

These hormones also promots glycogenolysis to increase the amount of blood sugar in the time of sudden demands. Vasoconstriction and vasodilation are also influenced by these hormones.

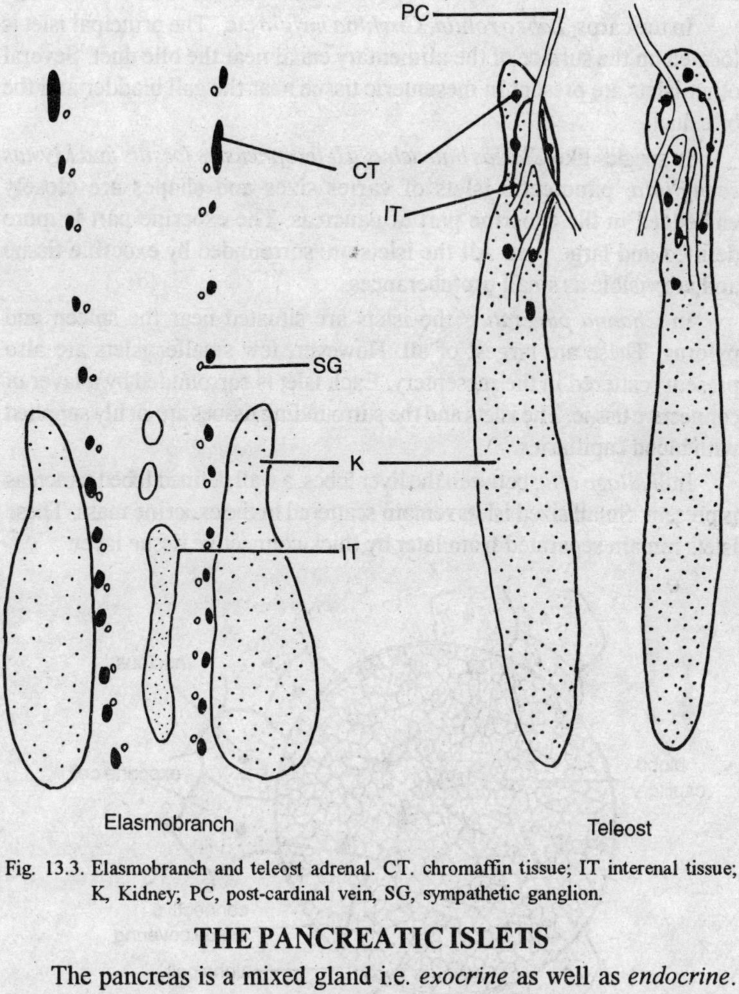

Fig. 13.3. Elasmobranch and teleost adrenal. CT. chromaffin tissue; IT interenal tissue; K, Kidney; PC, post-cardinal vein, SG, sympathetic ganglion.

THE PANCREATIC ISLETS

The pancreas is a mixed gland i.e. *exocrine* as well as *endocrine*. The endocrine part is the *islets of Langerhans*. The endocrine parts of pancreas are condensed into several islets which may have independent identity and may be scattered, in pancreatic mass. The islets are visible with the naked eyes due to their large size. These are called *principal islets*. They are generally situated near the gall-bladder, spleen or pyloric caecum. Their distribution and arrangement, however, varies in adults of different groups of fishes.

In cyclostomes (*Petromyzon*), the islets aggregate to form small masses in liver and intestinal wall. In *Myxine,* they form a thickening at the distal end of the bile duct.

In the carps, *Labeo rohita, Cirrhina mrigla* etc,. The principal islet is located on the surface of the alimentary canal near the bile duct. Several other islets are present in mesenteric tissue near the gall bladder and the bile duct.

In species like *Clarias batrachus, Heteropneustes fossilis* and *Mystus seenghala,* pancreetic islets of varies sizes and shapes are closely embedded in the exocrine part of pancreas. The exocrine part is more defined and large part. All the islets are surrounded by exocrine tissue and are visible as small protuberances.

In *Channa punctatus,* the islets are situated near the spleen and pylorus. These are largest of all. However, few smaller islets are also present scattered in the mesentery. Each islet is surrounded by a layer of connective tissue. The islets and the surrounding tissues are richly supplied with blood capillaries.

In *Wallago attu,* between the liver lobes, a wall defined lobed pancreas is present. Small sized islets remain scattered in the exocrine mass. These islets remain separated from later by thick connective tissue layer.

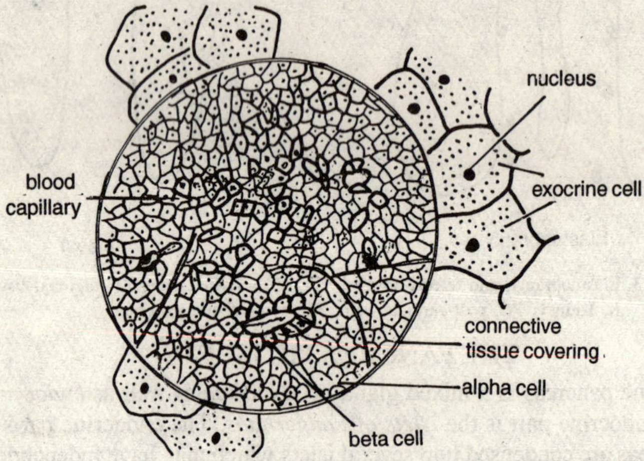

Fig. 13.4. T.S. Pancreatic Islet of a teleost.

In *Mostacembelus armatus* the principal islet is present in the duodenal loop at the anterior end of spleen and remains enclosed in thick connective tissue capsule.

Histophysiology of islet

In the islets of Langerhans three types of cells are identified. These are A or α-cell, B or β-cells and C-cells. The α-cells are roughty triangular in shape with eccentric nucleus and are aldehyde funchsin (AF) negative. The A-cells are of two types A_1 and A_2. The A_1 takes silver stain while A_2 is negative to silver stain. The A_1 cells believe to secrete *glucogon* whareas the function of A_2 is unknown. The number of A_1 cells is higher. The A_2 cells have been called as D cells by some workers. The B-cells generally concentrated towards the centre of islets. These cells are polyhedral in shape with a large round central nucleus. The B-cells are AF positive.

The B-cells secrete *insulin hormone*. The C-cells are agranular cells in the islets.

Insulin hormone is hypoglycemic factor brings down the concentration of sugar in blood plasma. The glucogon is hyperglycemic factor increases blood sugar level.

CORPUSCLES OF STANNIUS

These were first described by *Stannius* (1839) as discrete granular bodies in the kidney of sturgeon. These are small endocrinal bodies, that lie embedded in the posterior part or attached to the kidneys of ray-finned fishes. They are mistaken for interrenal body in section, but their embryonic origin in quite different, since they arise as evaginations from pronephric duct. They are generally oval or round in shape and vary in number from one to 10 in different species. In some fishes (gold fish, trout etc.) the corpuscles of Stannius are seen as flat, white structure on the peritoneal Surface of kidney.

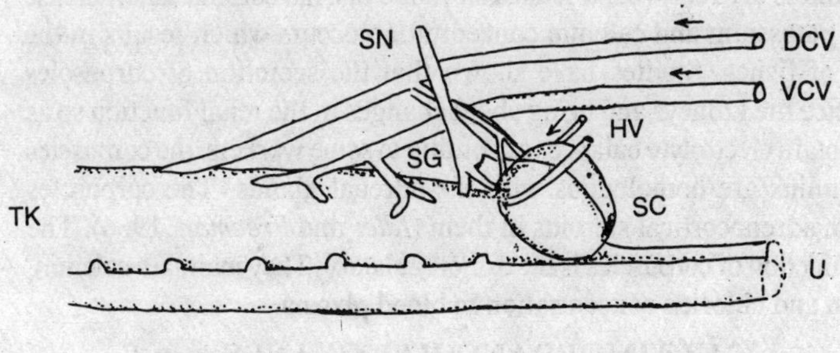

Fig. 13.5. Topography of Corpuscles of stannius in the three-spined stickleback, Gasterosteus aculeatus. SC, Stannius corpuscle; TK, trunk kidneys; SN, sympathetic nerve; SG, sympathetic ganglion sending branches to Stannius corpuscles and to kidneys; U, Ureter; DCV, dorso-caudal vein; VCV, ventro-caudal vein; HV, vein from hypaxial musculature.

Table showing number, size and location of corpuscles of Stannius in some teleosts (after *D.K. Belsare*, 1973)

Name of fish	*Number*	*Size in mm*	*Location from urinogenital aperture in mm*
Labeo rohita	2-3	3-6	50-60
Catla catla	2	2-5	30-40
Cirrhina mrigla	2-3	1-3	40-50
Clarias batrachus	2-10	0.5-1.2	4-30
Channa punctatus	1-2	0.2-0.8	10-20
Heteropneustes fossilis	2-4	0.4-0.9	20-25
Mystus vittatus	2-4	0.15-0.25	10-15
Notopterus notopteus	1	0.2-0.6	anterior portion of kidney

Histophysiology

The corpuscles of Stannius are composed of parenchymatous cells. These cells are arranged in the form of follicles or irregular cords. These are separated by connective tissue layer and surrounded by fibrous capsule. Within the parenchymatous cells or near them is present a ganglionic unit having blood capillaries and nerve fibres. Two types of cells are reported to be present in the corpuscles of stannius of salmon, (*Astyanx mexicanus* and *Anguilla agnuilla*). These cells are aldehyde fuchsin positive and negative. Only one type of cells has been reported by Belsare (1973) in several teleost fishes. The cells contain PAS-positive granules in the cytoplasm. Increase or decrease in the granules of the cells has been observed with changes in the salinity of water. Hypophysectomy also affects granulation. In the eel, *Anguilla anguilla* when the corpuscles of Stannius are removed, a reduction in the plasma sodium and increase in the potassium and calcium concentration occurs which results in the death of fishes. Studies have shown that the secretion of corpuscles influence the kidneys and bring about changes in the renal function so as to maintain electrolyte balance. According to some workers, the corpuscles of Stannius are homologous with the adrenal glands. The corpuscles contain adrenocortical steroids in them (*Idler* and *Freeman,* 1966). The main function of corpuscles is the osmoregulatory. They maintain calcium, sodium and chloride concentration in blood plasma.

ULTIMOBRANCHIAL GLANDS

The ultimobranchials arise as an epithelial evagination of last branchial pouch area in embryonic condition. Thus their origin is similar to the parathyroids. These are absent in cyclostomes. These are located on the transverse septum between the abdominal cavity and sinus venous just ventral to the oesophagus. The glands lie bilaterally, if single, lying in the middle line, however, in *Squalus acanthias* the gland lies unilaterally on the left side.

Histologically, the ultimobranchial gland comprises a large number of follicles. The epithelium of the follicle is made up of two types of cells, both the cells having membrane enclosed granules. In Osteichthyes the follicles are not seen and the gland becomes more or less diffused structure.

These glands are believed to be the source of calcium regulating hormone, *calcitonin*. *Calcitonin* controls hypercalcaemia by lowering the plasma calcium. Calcitonin is also concerned to migration and reproductive behaviours of fishes. However, definite role of this gland has to be ascertained.

GONADS

The *testes* in male and *ovaries* in female perform both, the cytogenic and endocrinal functions.

In male, there is one pair of testis. The testes are transparent when immature, on maturity they become creamy white. These are richly supplied by blood capillaries, particularly during breeding season. The testis is fully packed up with seminiferous tubules. In between the tubules are present *interstitial cells* or *Leydig cells.* These cells are also called lobule boundary cells. They represent the endocrine part of testis.

These cells normally appear during the breeding season and disappear afterwards in most of the fishes. The testes undergo cyclic changes throughout the year and there is variations in the appearance, weight and developmental stages of various spermetogenic cells.

The Leydig cells help in the biosynthesis of *androgens* from cholesterol (*Marhsall* and *Lofts*, 1956). *Testosterone* is the main androgenic hormone in fishes. In certain fishes (*Onchorhynchus nerka)* the testosterone is conjugated with glucuronic acid. Androgens in fishes is responsible for production of secondary sexual characters such as the colour pattern, modifications of fins and the expression of skin etc. These features are more prominant during breeding season.

In female, a pair of ovaries are present. These are covered by an ovarian wall. Between the wall and ovary there is a cavity called *ovocoel*. The source of ovarian hormone is *corpus luteum*. The corpora lutea are formed as a result of *follicular atresia* of advanced ooctye of maturing ovary. Each corpus luteum appears as a yellowish irregular mass of cells and produce *estrogen*. Cholesterol is the precursor of estrogen. It is present in the granulosa cells. Ovarian extract of pregnant sharks contains *relaxin* and in some fishes. the growth of ovipositor is controlled by *progesterone*.

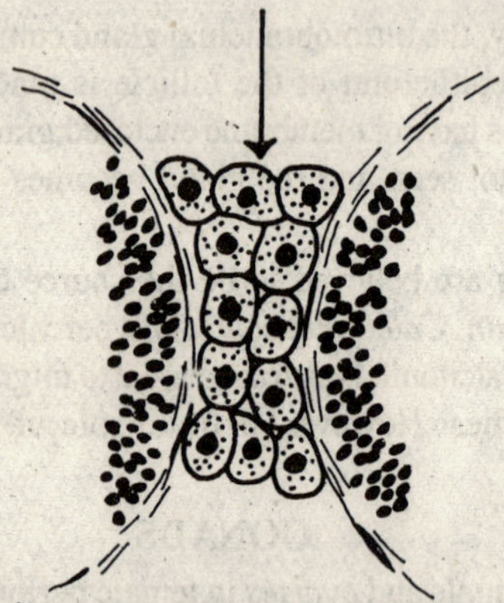

Typical vertebrate arrangement of endocrine intesrstitial Leyding cells

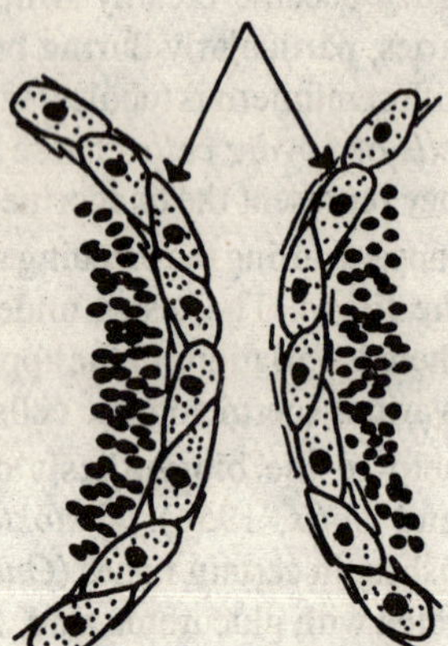

The Leyding cell homologue in the lobule walls of cortain teleost fishes

Fig. 13.6. Arrangement of Interstitial cells of Teleost fish.

PINEAL GLAND

The pineal gland or organ is present on the dorsal side of the brain and it is basically a *photoreceptor organ*, but it is also considered to be endocrine gland. The structure and function of pineal gland has been described by several investigators in recent years (*Boze*, 1961; *Kelly*, 1962;

Kappers, 1965; *Morita*, 1966; *Omura* and *Oguri*, 1969; *Hafeez*, 1971; *Srivastava* and *Dubey*, 1975; *Hafeez* and *Merhiga*, 1977; *Srivastava* et al, 1980; *Dubey* et al, 1982; *Khanna* et at, 1983; *Hore* and *Alim*, 1989 and others.

The pineal organ arises as a posterio-mid-dorsal evagination of epithalamus behind the habenula. It consists of a pineal sac, pineal thalamus and pineal stalk. In some fishes the pineal sac is bigger and touches the roof of the cranium while in others it is small. The sac is conical and elongate. It is highly vascular and is enclosed in a connective tissue capsule. In several species of fishes it is hollow, but may be solid in Lantern fishes. When the sac is hollow it is thick on all the sides or thin on the dorsal side and consists of sensory, supporting and ganglion cells. The supporting cells lie in between the sensory cells while the ganglion cells are found in the basal part of parenchyma. Scattered cell bodies can be seen in the lumen of pineal sac. The parenchyma of the sac protrudes as villi into its lumen and may form partitions in some cases. The lumen of the sac is connected with the diacoel through thalamus and stalk. The pineal stalk consists of low columnar epithelium. On the basis of several staining techniques the sensory cells have been differentialed into α-cell, β-cells, γ-cells, chromophobic cells, aldehyde fuchsin positive and negative cells. Pinealctomy results in reduction of growth rate and also some abnormalities appear in the skeleton. Recent experiments on *Mystus vittatus* has shown that pineal influences the maturation of gonads. Presence of *melatonin* and neurosecretory substance containing mucopolysaccharides, basic proteins, proteins containing arginine and lysine, acidic lipids, phospholipids etc. suggest that the pineal gland has difinitely some endocrine role which is still controversial.

THYMUS

The thymus gland is located on dorsal side or ventral side of the gill arches. It is an elongated and lobulated gland formed as solid ontgrowth from the dorsal angle of gill ponches. The gland is highly vascular and probably help in growth of the fishes.

UROHYPOPHYSIS

Fishes other than cyclostomes have a neurosecretory swelling at the posterior end of the spinal cord. It is called *urohypophysis* or *urophysis*. Its cells are similar to those present in hypothalamus. The neurosecretory cells are relatively large and have enlarged axons. The cells bodies sometimes called *Dahlgren cells* lie in the gray matter of the spinal cord. In teleosts, the axons terminate into small bulb-like structures resting upon the basement membrane of blood capillaries within the urohypophysis. Two biologically active peptides have been detected in the gland. These are named as *urotensin I* and *urotensin II*. The important functions of gland are osmoregulation, gas metabolism and buoyancy, smooth or involuntary muscle contraction etc.

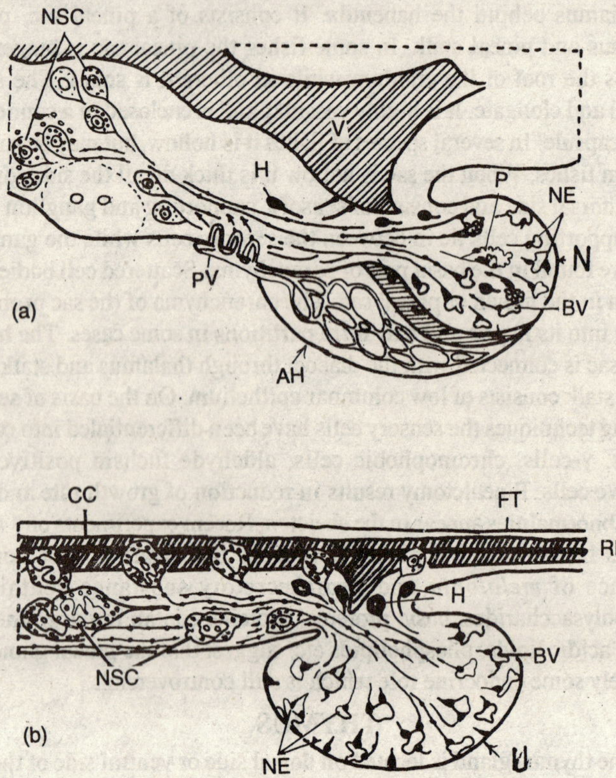

Fig. 13.7. Comparison of the Analagous pattern of organisation of the Caudal secretory system (b) with That of the Hypothalamohypophysial (a). AH, Adenohypophysis; BV, blood vessels; CC, central canal; FT, filum terminale; H, Herring body; N, neurohypophysis; P, pituicyte; PV, hypophysial portal vessel; RF, Reissner's fibre; U, Urophysis.

INTESTINAL MUCOSA

The intestinal mucosa produces hormone *secretin* or *pancreozymin*. Hormone secretin controls the production of pancreactic enzymes. This hormone was discovered for the first time by *Bayliss* and *Starling* (1902).

14

Poisonous and Venomous Fishes

At least 1000 species of fishes are known or thought to be venomous and poisonous. For the most part, these species are widely distributed throughout the world, and while their numbers may sometimes be quite large in some areas, they do not produce large ecological problems by virtue of their toxicity alone, nor at the present time are they other than a logical danger to men's health and economy. It is generally believed that most of the venomous and poisonous fishes have been identified, although it is conceded that a number of species have not yet been adequately described.

The term *poisonous fishes* is generally applied to those fishes whose tissues, either in part or in their entirely, are toxic.

The term *venomous fishes* on other hand, is usually limited to those piscines which are capable of producing a poison in a highly developed secretory organ or group of cells, and which can deliver this toxin during a biting or stinging act. In reality, all venomous animals are poisonous but not all poisonous animals are venomous. Fishes in which a definite venom apprature is present are sometimes called *phanerotoxic*, while fishes whose body tissues are toxic are called *cryptotoxic*.

Althoough the word venomous and poisonous are often used synonymously, most toxicologists, now try to confine the use of the term venomous fishes to those creatures having a gland or a group of highly specialized secretory cells, a venom duct, and a structure for delivering the venom while in the past there has been a tendency to employ the term *venom apparatus* to denote only the *sting* or *spine* used by the fish to deliver its venom, most investigators now use the term in its boader contaxt, that is, to denote the gland and duct in addition to the sting or spine. Poisonous fishes, as distinguished from venomous fishes, have no such apparatus, poisoining by these forms usually takes place through ingestion.

Fish poisoning is synonymous with *ichthyotoxism*. It is meant to exclude that type of poisoning which may occur following the ingestion of fish contaminated by bacterial pathogens. *Halsted* (1964, 1967) has suggested a further division of poisonous fishes.

1. Ichyosarcotoxic fishes

Those fishes which contain a toxin within their musculature, viscera, or skin.

2. Icthyootoxic fishes

Those fishes which produce a toxin generally confined to the gonads. In these piscines there is a relation between gonadal activity and production of toxin. Most members of this group are freshwater species. This subdivision would include those fishes whose roe are poisonous.

3. Icthyohaemotoxic fishes

Those fishes which have toxin in their blood. Some fresh water eels and several marine fishes make up this group.

4. Ichthyocrinotoxic fishes

Those fishes which produce a toxin by glandular secretion, but which lack a true venom apparatus. This group would include such piscine as the box fishes or trunk fishes, hagfishes and lamprey, all of which may produce a toxic substance in their skin and subsequently release their toxin into their environment.

VENOMOUS FISHES

Over 200 species of fishes are known or thought to be venomous. Among the more common of these pisces are the sting rays, weevers, scorpion fishes, zebra fishes, stone fishes, star gazer and certain of the sharks, cat fishes, rat fishes, and surgeon fishes. Venomous fishes are most frequently found in the Pacific areas, and for the greater part they are shallow water reef or inshore fishes. Some families, such as weevers, star gazer and sting rays, are chiefly benthonic and may sometimes be found in quite deep waters. Most venomous fishes are nonmigratory and slow swimming. They tend to live in a protected habitat in or around corals, rechs or kelp beds, or they spend much of their time buried in the sand.

Venom Apparatus

While it is not possible with in the confines of this chapter to discus the venom appraturs of every venomous fish, an attempt will be made to describe this structure in the most important or perhaps better known fishes. In most venomous piscines the venom apparatus is used as a defensive weapon. In some the use is not evident although fossil records would seem to indicate that even in these genera the sturcutre is also part of the animal's defensive armament.

Sting rays

The sting is composed of an inner core of vasodentine and the outer layer of enamel. It is bilaterally serrated and with the serration directed cephalically. There may be one or several stings in adult *Urolophus halleri*. The sting is approximately 4 cm. in length, while in larger species it may reach 40 cm in length. The spine is encased in an integumentary sheath. The venom is contained for the most part in the epidermal tissues within and overlying the two ventrolateral grooves of the spine.

In the stingray *U. halleri*, the integumentary sheath is composed of an inner variable layers of loose areolar connective tissue covered by an outer layer of epithelium. The inner dermal layer is rich in thin walled blood vessels. Occasionally, it projects into the epidermis, forming delicate papillae which carry nutritive blood vessels. The epidermis is composed of basal group of flattened, heavily pigmented cells. In places these cells form a distinct pigmented layer. The innermost layer of these basal cells is perpendicularly oriented. As these cells mature, their nuclei become smaller and stain more intensely. A layer of varying thickeness of oval, vacuolated cells, having nucleus compressed at one end and eosinophilic cytoplasm, over lies the basal epithelium in many areas. These cells occupy 20 to 60% of the total tissue area of the epidermis in and over the ventrolateral grooves. The cells have been called *glandular* by *Evans* and *Halstead* and *Bunker*. They give positive reactions to several protein tests and are believed to contain venom. In some areas, the narrow columns of modified squamous cells connect the basal layers of the epidermis with the outer layers. An outer layer of partially cornified cells with small dark staining nuclei covers the vacuolated and squamous cells. On the surface is a thin layer of dense, cornified material in which no cell detail remains.

The discruption, however, is quite different from that detailed by *Porta* (1905), *Evans* (1916) and *Fleury* (1950) who described the glandular cells as being in the inner or dermal layer of the integumentary sheath, the area we have described as containing the loose areolar connective tissues and blood vessels. *Fleury* states that in *Myliobatis aquila* the venom is evaculated through one or two excretory canals as their ranufications in the interdental species. The ducts described by *Fleury* are very similar to those we have noted as blood vessels. Many of our sections show blood cells in these structures. It is not particularly clear from the Fleury's paper where these ducts arise and where they terminate.

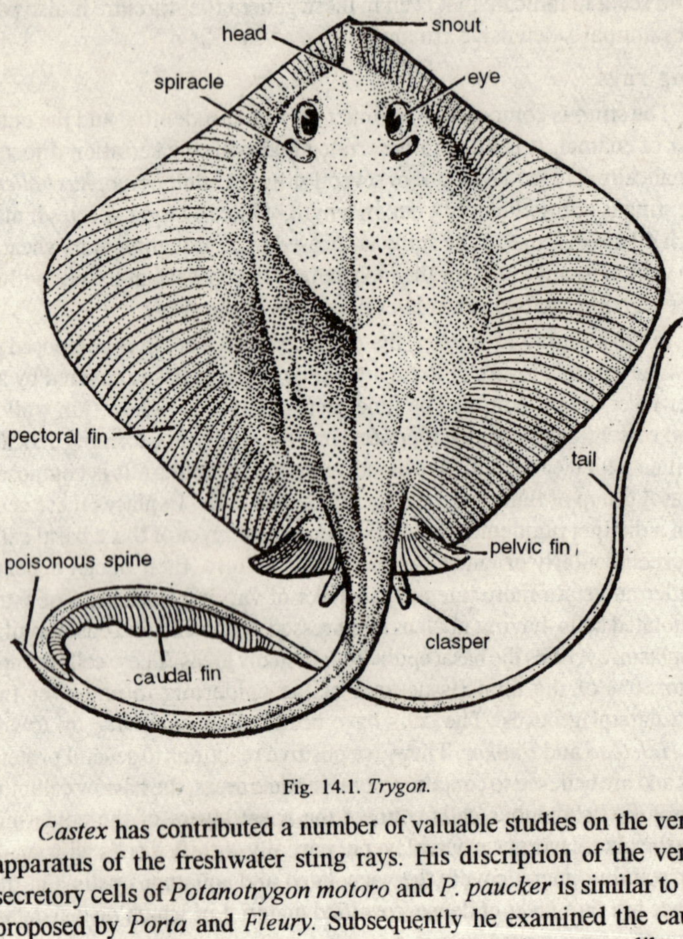

Fig. 14.1. *Trygon.*

Castex has contributed a number of valuable studies on the venom apparatus of the freshwater sting rays. His discription of the venom secretory cells of *Potanotrygon motoro* and *P. paucker* is similar to that proposed by *Porta* and *Fleury.* Subsequently he examined the caudal spine of *P. magdalenae* and presented a description not unlike that presented by *Ocampo et. al.*, (1953) and by *Russel* and *Lewis* (1956). More recently he has summarized the findings of his studies and suggests that in the river sting rays the venom cells are probably in the epidermis.

Pearson (1967) has examined the stings of seven speices of sting rays found in Australian waters. In all species the venom producing cells were located in the epidermis of the integumentary sheath. No venom producing cells are found in tissues of the dermal layer within the ventrolateral grooves, which *Pearson* calls the *dermal triangles* a term this writer feels is preferable to the commonly employed glandular triangle *Pearson* describes the venom producing cells as *club cells*, located for the most part in the epidermis of the ventrolateral grooves of the spine, although, as described by previous workers some are also found in the dorsolateral and ventrolateral epidermis enveloping the spine.

The club cells were found to contain large amount of proteinaceous materials. Their cytoplasm was either homogenous or slightly granular, except for a single spherical to rod shaped cells inclusion and the nucleus. The cytoplasm of the club shaped cells was eosinophilic. Tests for mucous substances were negative, they were weakly positive for lipids, and strongly positive for proteins.

Stingings by sting rays occur frequently in humans, but apparently infrequently in marine animals. Fossil records, however, would seem to indicate that at one time a number of marine animals, particularly shortly fed upon sting rays and were apparently stung by them. Recently we found the caudal spine of two small sting rays embedded in the linning of the stomach of a young seal.

Weevers

The weevers are small marine fishes of the family Trachinidae. The venom apparatus of the weevers consists of 2 opercula spines, 5 to 8 dorsal spines, and the tissues contained within the integumentary sheath surrounding each spine. In the general weever, *Trachinus draco*, the two dentinal opercular spines extend posteriorly and very slightly downward from near the superior margin of each operculum. Each spine is firmly attached to the operculum for the proximal one third of its length, and lies free and superficial to the posterior portion for the distal two third. These spines are covered by a thin walled integumentary sheath which, when removed, exposes the white, soft, spongy glandular tissue along the superior and inferior margins. A groove exists along the inferior and superior margins. Within the superior and inferior grooves, and the two conccavities, into which they pass at the base of the spine, lies most of the spongy glandular tissue that produces venom. The five to eight dorsal spines are enclosed within individual integument sheaths connected by interspinous membranes. In a 38 cm *T. draco*, the first dorsal spine measures approximately 21 mm. Cross sections through the opercular spine near the operculum reveals the peripheral layers of integument continuous with an inner or dentinal layer, and a glandular area between the integument is composed of an epidermis of stratified squamous epithilium wihich rests upon a basement memberane which in turn seperates the epidermis from the dermis. Large secretory or mucous cells with clear cytoplasm are seen scattered through out the epidermis. The dermis is composed of a thick layer of dense fibrous connective tissue and is seperated from the spine by a space. Projecting into each of the glandular grooves of the spine is a thin layer of dermis and glandular tissue from the glandular areas, which appears embedded in the surrounding dermis. The glandular parts are made up of masses of large polygonal distended cells of variable sizes, staining light pink and filled with finely granular secretions.

The integumentary sheath of a dorsal spine is composed of an epidermis of stratified squamous epithilium of five or six cells thick resting on a thick pigmented membrane, a dermis of dense, fibrous connective tissue, and the gland tissue with in the anterolateral glandular grooves. The gland cells are large, polygonal cells of varying sizes, perpendicularly oriented to the axis of spine, and staining pink. Their cytoplasm is finely granular. In both the opercular and dorsal spines envenomation takes place when the integumentary sheath is ruptured by trauma.

Star-gazers

The star-gazers are small, bottom dwelling marine fishes of the family Uranoscopidae. The venom apparatus of the star-gazer *Uranoscopus scapes* consists of the two cleithral or shoulder spines, their venom glands, and the enveloping integumentary sheaths. The spines are sharp, conical, acellular structures of cementum-like materal having concentric growth laminae. They protrude from the superior posterior edge of the operculum above the pectoral fin. The distal portion of each spine is almost round, while the proximal portion tends to be flattened in the medial lateral plane. Their lengths in a fish measuring 170 cm would be approximately 10 mm. Along the inferior edge is a shallow groove which is lost toward the middle of the spine. A similar, but much less marked groove is found along the superior edge.

The integumentary sheath enveloping the spine has an epidermis composed of a basal layer of columnar cells, stratified squamous epithelium having cuboidal or polyhedral cells interspersed with unicellular mucous glands. The dermis is seperated from the epidermis by a prominent basement membrane. It is relatively thick, moderately dense, and contains spaces filled with a geletinous material. The geletinious material is believed to contain the venom. Areas of this material are often associated with fibrocytes, although the association is not clearly definable. *Halstead* and *Dalgleish* noted that in the large spaces within the spine itself, they also found a gelatinous material, as well as thin walled blood vessels. They suggest that modified fibrocytes either produce the venom continuously, or hypertrophied fibrocytes produce a holocrine secretion which is collected within dermal pockets or cancellous spaces. Passage of the venom to the tip of the spines is accomplished by means of the shallow groove extending along the inferior edge of the spine when pressure is put upon the sheath, as during the stinging act, the venom containing cells as well as the sheath rupture and the venom is forced into the groove of the spine and then to the tip.

SCORPION FISHES

(a) Notesthes robusta

This scorpion fish is commonly known as the bull rout, occurs through out the estuarine and coastal river systems of Eastern Australia, particularly in shallow, weed infested creeks. Its venom apparatus consists of the 15 dorsal spines, 3 anal spines, 2 pelvic spines and the gland and tissue associated with these spines. The first twelve dorsal spines are strongly curved caudally. The remaining ones are only slightly curved. The tips of the dorsal spines are normally exposed when the fin is errect. Both the anal and pelvic spines are directed caudally. Each spine is enclosed in an integumentary sheath which is continuous with the web of the fin.

Removal of the integumentary sheath reveals a fusiform venom gland in the paired anterolateral grooves of the spine. These venom glands are straw coloured and show numerous opaque areas when the gland is in the fresh state. They terminate 1-2 mm. from the tip of the spine. The mean ratio of gland length to spine length is approximately 1:3.

A cross section through a typical dorsal spine reveals an epidermis composed of a basement membrane, a basal layer of cells, a layer of stratified epithilium five to ten cells thick, and an outer single row of columnar cells possessing finger-like outgrowths. This epidermis is approximately 60μ in thickness. Within the epidermis, two types of gland cells are found. One is a spherical type with a small peripheral nucleus and a large vacuole, the contents of which are non-granular. These cells are fairly constant in size, averaging 16μ in diameter when found at the outside edge of the epidermis, they appear as goblet cells. The second type of gland cell is oval to circular in shape, it possesses a peripheral nucleus and a granular cytoplasm when circular, these cells average 26μ in diameter. The underlying dermis is four to five time thicker than the epidermis and is composed chiefly of collagenous connective tissue.

The two venom glands in the groove of the spine are closely invested in a capsule of collagenous connective tissue, which also invest the spine itself. The glands are pear-shaped, circular or oval, when circular they measure approximately 350μ in diameter. A sheet of cells identical in appearance and staining qualities to a stratified cells of that epidermal is found across the center of the gland. On either side of this sheet is a single row of very large, rectangular gland cells.

Between the gland cells, and particularly in the periphery of the gland occasional granule filled sacs of approximately 60μ in diameter are found. The granules are spherical and average 1.5μ in diameter. Further study of these cells and sacs by *Cameron* and *Endean* revealed that the gland cells became the granule filled sacs, and that these sacs tend to push toward the periphery of the gland. There does not appear to be any duct from these glands either to the tip of the spine or to the base. Expulsion of the venom takes place when the integumentary sheath is torn or retracted along the spine in the direction of the base.

(b) Pterois volitans

The zebra fish, lion fish, or turkey fish is a scarpaenid usually found in shallow waters around corsal heads, in underwater caves, and about under water debris. Because of its brilliant colours and long delicate fins, it is easily seen and rarely contracted accidently. Its venom apparatus consists of 13 dorsal spines, 3 anal spines, 2 pelvic spines, the envoloping integumentary sheaths, and the glandular complex lying with in the anterolateral grooves of the spines.

The dorsal spines are almost straight, except at the base and the tip where they incline slightly caudally. The anterolateral grooves originate just above the base of the spine and extend the entire length of the structure. In a 45 mm spine the groove may measure 39 mm. The spine is invested in the integumentary sheath, a thin layer of spectacularly coloured, fibrous connective tissue when the anterolateral grooves are fusiform shaped venom glands, which may occupy three quarters of the total length of the spine. The glandular structures with in the grooves are composed of several types of tissues, including the venom producing cells, which are large, polygonal shaped cells with pinkish gray, finely granular cytoplasm. According to *Halstead et. al.* (1955) these cells may measure 207μ by 75μ or more and are collected together in connective tissue compartment within the anterolateral grooves. They are covered in turn by the integumentary sheath. There does not appear to be any glandular duct through which the venom is secreted or discharged. Envenometion probably occurs through mechanical pressure on the spine and rupture of the sheath.

(c) Scorpaena guttata

The California sculpin, or scorpion fish, is a common inhabitat of the rocky shores and relp beds of the southern California coast. It is involved in numerous stingings to fisherman. Its venom apparatus consists of 12 dorsal spines, 3 anal spines, 2 pelvic spines and their enveloping integumentary sheath. The spines are shorter and heavier than in *Pterois*, and the sheath in thicker.

When the integumentary sheath is removed from a drosal spine, a slender, elongated, fusiform strand of grayish tissue is found lying with in the distal one half or two third of the glandular grooves on either side of the sting. The venom is contained with in this tissue. The venom glands are similar to those found in *Pterois*, although more highly developed. Envenomation occur through mechanical pressure on the spine, which generally tears the integumentary sheath.

(d) Synanceja herrida

There are a number of species of *Synanceja*, and they vary in length as adults from 22 tc 48 cm. They inherit their name from the fact that they so remarkably resemble stone or coral. They have the habit of lying motionless in coral or rock, or partially burried in the sand or mud, and so still they lie that molluscs and other benthonic marine animals may often crowl over them. According to *Endean*, if approached, while they are swimming, they will frequently turn themselves so that their dorsal spines point toward the intruder.

The venom apparatus of *Synanceja horrida* consists of 13 dorsal spines 3 anal spines, 2 pelvic spines and thick, warty integumentary sheaths enveloping these spines, the two glandular grooves in each spine and the venom glands and their ducts within these grooves.

They dorsal spines are short, stout, and straight. The two anterolateral glandular grooves extend almost the entire length of the spine. The glands appear as two large fusiform masses at the middle and toward the tip of the spine. The distal end of each gland terminates in a duct like structure lying within the grooves. These ducts extend from the glands to the tip of the spine.

POISONOUS FISHES

A few of the more important works on the poisonous fishes and a discussion of the data on the chemistry and biological effects of these interesting toxins are given below.

Excluding those fishes which become toxic because of bacterial pathogens or the products produced by them, or by the process of tissue decay, approximately 700 species of fishes are known to be poisonous. Most of these species are found in the coral reef belt. They tend to occur in greater numbers around islands than along continental shores, although their distribution is spotty even in a particular part of occean or around an island. Most poisonous fishes are non migratory reef fishes. They may be either herbvores or carnivores. Some species have tissues which are toxic at all times. Other speices are poisonous only at certain periods, or in certain area, while still others have only specific tissues or organs which are toxic, and the toxicity of these vary with time and place.

Early writers were inclined to attribute fish poisoning to a single common factor. Thus one finds that numerous causes for ichthyotoxism are discribed in the early literature. Among the more often mentioned are :

1. Feeding on toxic or non-toxic marine plants. In the latter case it was suggested that the metabolic processes of the fish were able to alter certain plant components into a toxic form.
2. Feeding on toxic protistan, particularly dinoflagellates.
3. Feeding on shoreline terrestial plants whose berries or leaves have been swept into the sea.
4. Pollution of the ocean by industrial poisons.
5. The dumping of large quantities of metals, particularly copper, into the ocean, such as occured in the Pacific at the end of World Was II.
6. Epidermis in fishes, giving rise to toxic endogenous substances.
7. Spawning activities
8. Feeding on corals
9. Feeding on jellyfish
10. Feeding on Palolo worms.
11. Feeding on toxic molluscs.
12. Feeding on Zooanthellae in corals.
13. Feeding on other toxic fishes.
14. The dumping of radioative materials into the ocean.
15. Climatic changes.
16. Leaving the fish exposed to moonlight.

It is not the purpose of this chapter to review or discuss the factores which contribute to ichthyotoxin. The reader is referred elsewhere for a detailed survey of this problem.

(A) Ichthyosarcotoxism

This type of poisoning is caused by ingestion of fishes containing a poison within their musculature, viscera, skin or body slime. It is generally identified with the kind of fish involved cyclostome, elasmobranch, *Chimaera*, *Ciguatera Tetraodon*, scombroid, clupeoid and gempylid

Hallucinatory fish poisoning is also identified with this type of poisoning. The toxin is distinctly defferent in some of these fishes (*Ciguatera* and *tetradon*), while it is the same or very similar in others (*Cigulatera*, clumpeoid).

Cyclostome poisoning

This vague entity has been discribed in the early literature, but it is difficult to seperate the clinical cases of possible cyclostome poisoning due to the ingestion of lampreys or hagfishes from gastrointestinal disturbances precipitated by the bacterial contamination of these fishes. *Prokhoroff* (1884) stated that cyclostome poisoning from the eating of lampreys occured frequently in the Yambu District of Russia. Other references of poisoning by these fishes are recorded by *Halstead* (1967). The slime of the hagfishes is reported to be toxic and some authors have attributed the toxicity of these fishes to slime which has been improperly removed prior to eating.

Nothing is known of the chemical and pharmacological nature of the suspected toxin from cyclostomes. It is thought by some to be an amine but this seems unlikely. No definitive study on the skin, musculature or slime of thse fishes has been carried out. The buccal secretions of certain lampreys contain anticoagulant.

Elasmobranch poisoning

As with cyclostome poisoning, intoxication following the ingestion of shark flesh has been noted in the early literature but again it is difficult to distinguish between true cases of ichthyotoxism and those of food poisoning. *Coutaud* (1879) descreted seven cases of poisoning following the eating of the liver of an unidentified shark off. New Caledonia in 1873. *Cooper* also noted that the liver of the tiger shark, white tipped lagoon shark, and perhaps other large sharks may be toxic and produce symptoms and signs not unlike those of *Ciguatera* poisoning, although perhaps more severe and of quicker onset. Deaths from eating shark liver have been reported as recently as 1957, 1960 and 1961 in the Gilbert Islands.

Jensen has reported on intoxication in dogs and man following the eating of fresh shark flesh in Greenland. In lethal poisonings the agonal period was followed by tonic and clonic convulsions and death.

Hastead and *Schall* (1958) found the liver and gonads of the gray reef shark *Garcharhinus manisorrah* of the line islands to be toxic on intraperitoneal injection to mice. *Li* (1965) stated that extracts of the liver of the gray shark provoked similar pharmacological responses to those produced by extracts from the red snapper but *Janus bohar*. He concluded that the toxic material was a cholinesterase inhibitor. This conclusion rests upon questionable deductions and experimental data which are not easy to interpret.

Several skates, rays and rat fishes have also been implicated in poisoning to man, but again the experimental and clinical pictures are not clear. It is possible that in the species of skates and rays found in the tropical waters, the toxin, which appears to be present in greater quantities in the viscera than elsewhere, may be similar to shark liver toxin, and may be related to cignatoxin. Aqueous extracts of the oviducts of large specimen of the *Chimaera, Hydrolagus colliei* were found to be toxic by Halstead and Bunker, but nothing is known of the chemical or toxicological properties of Chimaera poison.

It is obvious that elasmobranch poisoning is a complex entity and that the etiological factors which contribute to this intoxication may be appears to be different from that found in the flesh of sharks from certain tropical islands. Skates and ray poisoning in the tropical and subtropical oceans appears to produce similar clinical manifestations as shark liver poisoning and the two may be related to *Ciguatera* poisoning *Chimaera* poisoning seems to be a completely different entity.

Ciguatera poisoning

The term *Ciguatera poisoning* is applied to a type of intoxication characterized by certain gastrointestinal neurological manifestations. It may occur following the ingestion of any of a considerable number of tropical reef and semipelagic marine fishes, including the barracudas, groupers, sea basses, snappers and eels. *Halstead* lists over 440 species of marine fishes which have been implicated as being Ciguatoxic. This type of poisoning may also be found in certain marine echinoderms, molluscs and arthorpods since almost all of the fishes noted above are normally edible, and many are valuable food fishes in some parts of the world, *Ciguatera poisoning* is not only the most common but also the most treacherous forms of ichthyotoxism.

The majority of these fishes implicated in ciguatera poisoning are reef or shore species, a few are open water forms. Almost all of the toxic species are found between latitude 35° N and 34°S. They are usually bottom dwellers although seldom found below depths of 200 feet. They are more likely to be found in lagoons than along seaward reefs. Most toxic species feed on benthic algae, benthic fishes, or some other benthic organisms. According to *Randall* (1958), none appear to be a positive corelation between the amount of fish in the diet and the degree of toxicity.

There is also a tendency for the larger fish of a speices to be more toxic than the smaller fish of the same species. In most cases the flesh is less toxic than the viscera. The liver is usually the most toxic part of the animal, although the reproductive organs may be equally as poisonous.

The poison probably originates in a benthonic orgnaism. Most investigators feel that the source of toxin in marine algae. Algae which have been implicated are *Lyngbya majuscula*. Other Cynophytae may no doubt sometimes be involved and indeed perhaps other algae. It is though that the poison in the algae or some product of the algae is trasvectored from the algae to herbivorous fish, which in turn is eaten by man, or that the herbivorous fish which is turn is eaten by man, or that the herbivorous fish is eaten by a carnivorous species which is subsequently eaten by man. It may be that the end poison is a complex of toxins, or perhaps there are several toxins that can be present singly or in combinations. Since certain marine invertebrates are also known to contain the toxin, or a substance very similar to the toxin, it may be that in some cases the poison is transvectored through these animals to carnivorous fishes and hence to man. It is also possible that zooxanthellae may be a source of the poison.

It is known that non toxic fish fed a diet of toxic lutianus bohar may become toxic.

Tetrodon poisoning

Tetrodotoxin (TTX), puffa or fugu poison is found in certain puffers, ocean sun fishes, tigger fishes. *Halstead* lists over 75 teterodotoxic fishes reported to be poisonous. TTX is also found in certain amphibians species

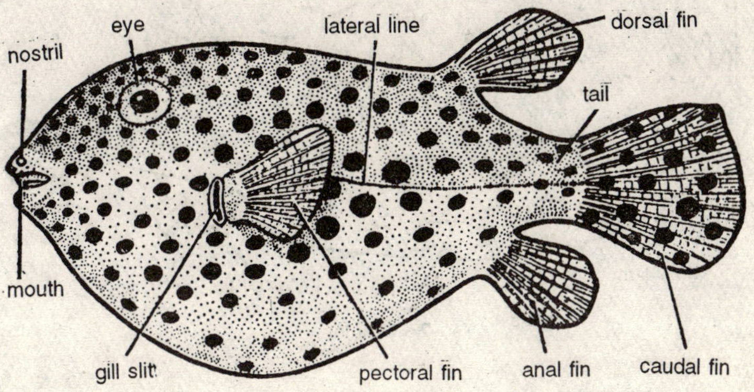

Fig. 14.2. *Tetrodon.*

of the family Salamandriadae. The puffers or puffer like fishes appear to be the only piscines unversally regarded as poisonous.

In most cases the toxin is concentrate in the ovaries and liver, with lesser amounts being found in the intestine and skin and only small amounts in the body musculature and blood. In almost all fish species so far studied, the concentration in the ovaries has been considerably higher than in the corresponding male tissues. The appearance and amount of toxin in the fish is related to the reproductive cycle and appears to be greatest just prior to spawning which in turn varies with the species involved and the locate.

Scombroid poisoning

Certain of the mackeral-like fishes, the tunas skip jacks, and fonitos and the japonese saury, *Cololabis saira*, are occassionally involved in poisoning to man while the symptoms and signs of scombroid poisoning in man are quite different from those produced by ciguatera poisoning, these fishes can also be involved in ciguatera poisoning. If scromtoids are inadequately preserved, a toxic substance is said to be formed with in the body musculature. This substance was once thought to be histamine formed by the action of enzymes and bacteria the death of fish. However more recent evidence would seem to indicate that the toxic component may not be histamine although the complex does resemble this amine in certain of its properties. *Kawabala* has given the toxic substance the name sourine.

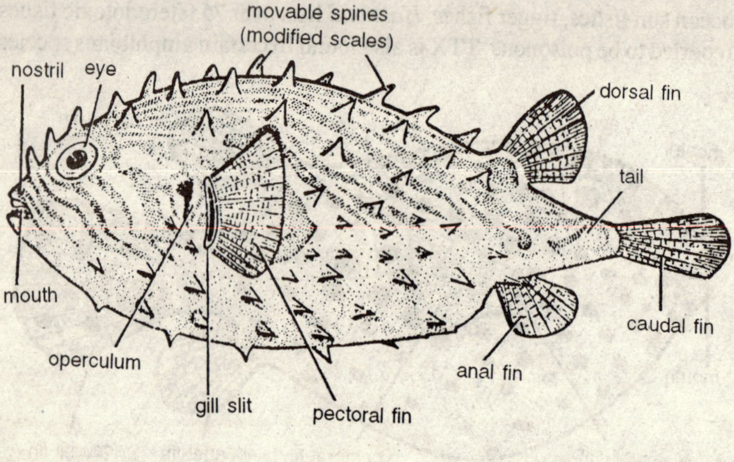

Fig. 14.3. *Diodon.*

Sourine is said to closely resemble acetylcholine and histamine in some of its pharmacological properties, but it is chemically distinct. It has an Rf value of 0.1 with n-butanol, acetic acid and water (4:1:2) and is insoluble in ether, acetone, benzene, and chloroform. It is dialyzable, can be precipitated with phosphotungstic acid, and has certain other properties which would indicate that it has relatively low molecular weight. It appears that the toxic complex of scomboid poisoning may be saurine and histamine, and perhaps a third substance.

Clupeoid poisoning

This form of ichthyosarcotoxism may occur following the ingestion of certain herrings and tarpons. Since most of these fishes are valuable food fishes, this entity is one of considerable concern in certain parts of the world. *Helfrich* notes that one species of herring is thought to be toxic at all times in Fiji, while a second species is poisonous only during October and November.

The biogenesis of the toxicity is not known. It would seen that toxicity is associated with the food chain, but the contributing organism has not been identified. It has been suggested that a dinoflagellate, or a blue green algae or even the palolo worm may be involved.

Nothing appears to be known about the chemistry and toxicology of this poison. In 1967, several reportedly toxic *Clupea venenosa* were sent to laboratory for examination. The viscera and some flesh was fed to a cat. The cat smelled but did not touch the viscera. It did eat a small portion of the flesh before abruptly stopping and retreating to a cornor, where in about 4 min it asumed a most un-cat like stance, with all extremities extented and its head resting on the floor. It lay on its abdomen, panted with unusuall force and retched several times. The pupils were dialated, deep relfexes seemed normal but the cat could not be made to stand the righting reflex was intact. At approximately 7 min following the meal the cat began to salivate profusely to vocalize, and it assumed a guarded position. The signs subsided over the next 2 hr, and other than passing two loose stools there were no further significant changes.

Gempylid poisoning

The flesh of certain fishes, particularly the gempylids, is known to have a high concentration of wax. The chemistry and physiopharmacology of the musculature of the castor oil fishes *Lepidocybium flavobrunneum* and *Ruvettus pretiosus cocco*, both belonging to the family Gempylidae, have been studied by japnese workers. *Kimura* found that cetyl and oleyl alcohol were the cheif component of the unsaponifable fraction from

R.pretiosus. *Cox* and *Reid* classified the oil of the flesh as a liquid was consisting mainly of cetyl and oleyl esters of oleic and hydroxy oleic acids. *Matsumoto et. al.* reported that the oil of the fish of *L. flavobrunneum* was also composed of cetyl and oleyl alcohol. *Mori et. al.* analyzed the flesh of both species using modern chromatographic techniques. The wax, which made up 20% of the acelone soluble oil, contained cetyl and oleyl alcohol and oelic and eicosenol acids.

The castor oil fishes are not usually used as food. Their seborrheic and diarrhoeal effects in mammals are well known. Animals maintained on the diet of castor oil fish lose weight and eventually die.

One of the cod, *Erilepsis zonifer lockington*, also provokes diarrhoea when eaten in quantity. *Kaneda* has shown that the oil of the flesh is composed cheifly of glycerides, high in olic acid. The toxin appears to be different from that found in castor oil fishes.

(b) Ichthyocrinotoxin

Ostracitoxin and Pahutoxin

Certain fishes are known to release a toxic substance from the skin which is capable to killing other fishes and perhaps other marine animals. This toxin appears to be part of the animals defensive armament and is perhaps released as an alarm substance to deter possible predators.

15

Larvicidal Fishes

Larvae of carnivovous fishes feed upon the zooplanktons (cladocerans and rotifers) where that of herbivorous fishes feed upon phytoplanktons (algae, aquatic weeds etc.). *Atkin* asserted that there are some fishes which consume the insect larvae as their favourite food. Mosquitoes are the vector of several diseases such as malaria, yellow fever, dengue and filaria. These diseases can be eradicated with the help of these larvicidal fishes by gradually decreasing mosquite population. This is known as *biological control*. Biological control now regarded as one of the most potent tool to destroying mosquitoes without causing any harmful effect on human health and food chain.

Characteristics of larvicidal fishes

A larvicidal fish, introduced to control mosquitoes, should have following features;

1. The fish should be of small size so as to be capable of moving freely among the weeds in hunting larvae.
2. These fishes should not be so delicate rather they should be hardy with respect to the ecological conditions of their habitat and transportation.
3. It should have no food value.
4. Its movement should be swift enough to escape from its enemies.
5. It should be carnivorous fish and its main diet should be insect larvae.
6. It should be surface feeder.
7. It should be capable of free breeding.

Important larvicidal fishes

These are as follow :

(A) ***Exotic species.*** The exotic species of larvicidal fishes are : *Gambusia affinis, Lebistes reticulatus* and *Carassius carassius.*

(B) ***Indigenous Species.*** *Notopterus, Oryzias, Barillus, Rasbora, Puntius, Aplocheilus, Esomus, Anabas, Colisa, Etroplus, Mugil, Anabas, Colisa, Ambassis, Theropon, Haraichthys etc.* are some indigenous sepcies of larvicidal fishes.

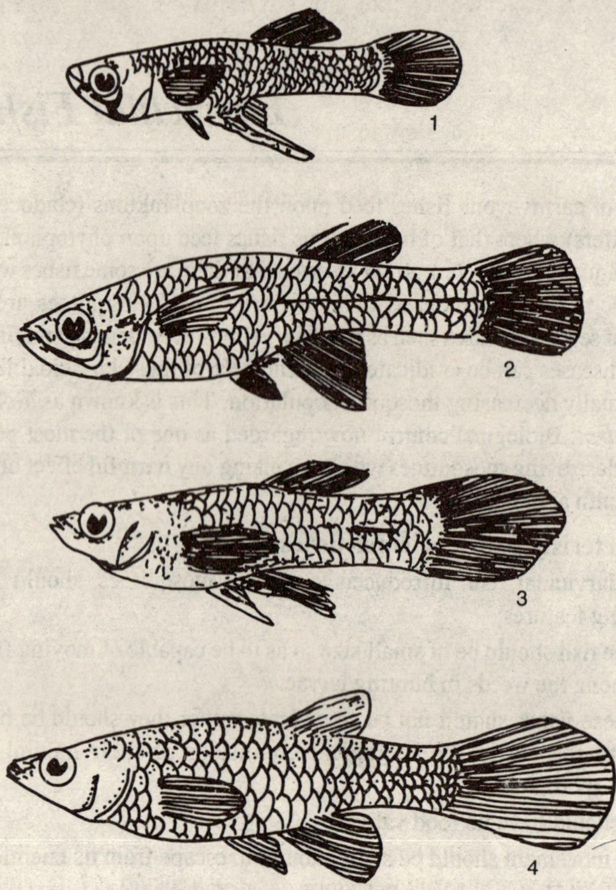

Fig. 15.1. Larvivorous fishes (exotic). 1. *Gambusia affinis* (male), 2. *Gambusia affinis* (female), 3. *Lebistes reticulatus* (male), 4. *Lebistes reticulatus* (female).

(A) ***Exotic species.*** Of the exotic species, *Gambusia affinis (*Top minnow) is used extensively for mosquito control in several species. It is only 2 $^1/_2$- 3" in length. Certain species of *Gambuzia (G. affinus holbrooki)* can be kept in the aquarium. Maturity in this fish is reached only after four months. It has been reported that this fish consumes 165 larvae within 12 hrs.

Carassius auratus (gold fish) has been imported to various countries as the ornamental fish but due to its larvae feeding habit it can be used for mosquito control. It has high cost that in why it can not be used on a lage scale.

Lebistes has also been brought to India and it is found in the ponds and tanks of S. India.

(B) ***Indigenous Species.*** *Notopterus notopterus :* It occurs in fresh and brakish water of India. The young stage of this fish feeds upon the insect larvae but the adult feed upon small fishes. As the fish consumes other small fishes, it can not be recommended for antimalaria use.

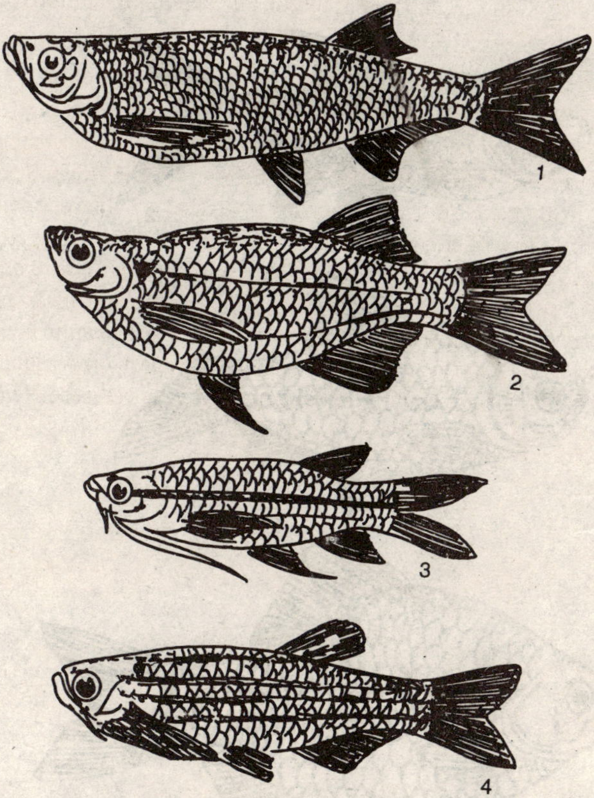

Fig. 15.2. Larvivorous fishes. 1. *Oxygaster bacaila,* 2. *Laubuca laubuca,* 3. *Esomus danricus,* 4. *Danio rerio.*

Aplocheilus lineatum. It is commonly known as Indian Top minnow. It is found in fresh and brackish waters. It grows to a lenth of 9 cm. It is a hardy fish and an effective larvicidal fish.

Oryzias melanostigma : It is a fresh water fish. It breeds in shallow water. The fish is very useful as larvicidal form but delicate and can not be transportes easily.

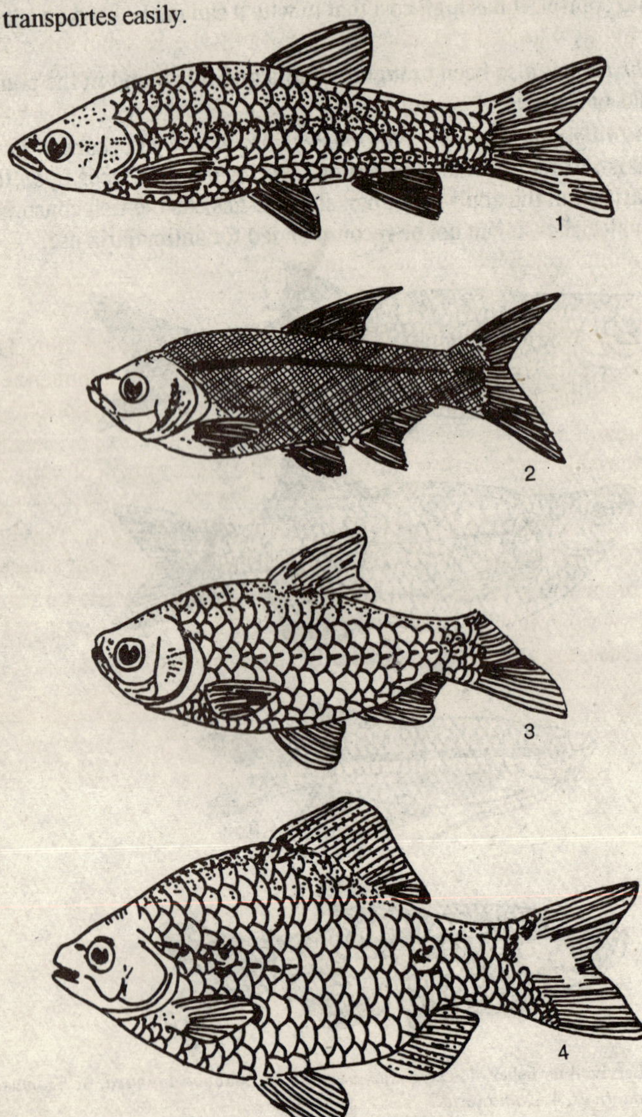

Fig. 15.3. Larvivorous fishes. 1. *Rasbora daniconius*, 2. *Amblypharyngodon mola*, 3. *Puntius sophore*, 4. *Puntius ticto*.

Aphanius dispar : A single species occurs in India. It has been found to be an effective larvicidal fish.

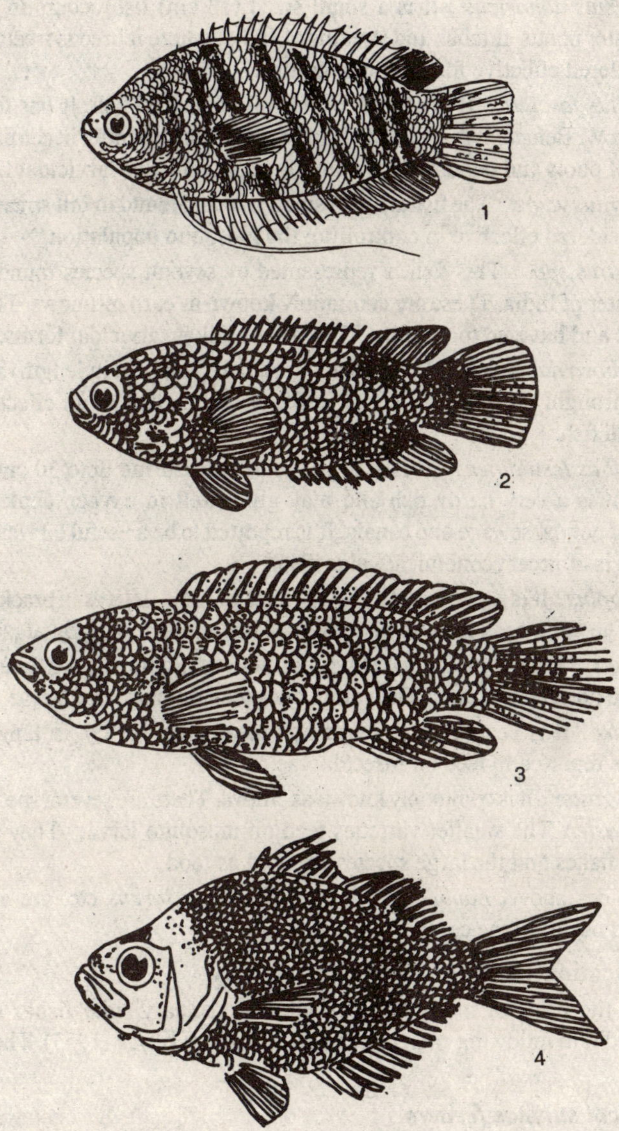

Fig. 15.4. Larvivorous fishes. 1. *Colisa fasciatus*. 2. *Anabas testudineus*, 3. *Badis badis*. 4. *Chanda ranga*.

Danio rerio : It is a small sized (5 cm in length) fish with upturned mouth and has some value in mosquito control.

Esomus danaricus : It is a small sized (13 cm) fish occurs in the fresh water ponds, ditches and streams of India, where it breeds freely. It is considered effective for mosquito control.

Colias fasciata : It is a smal, beautifully coloured fish. It has food valvue in W. Bengal. It is a hardy fish and capable for spawning in confined waters of pools and ponds. It has found to be an effective larvicidal fish.

Barilius vagra : The fishes are of small size and found in hill streams. It is considered effective in controlling the mosquito population.

Puntius spp. : This fish is represented by several species found in fresh water of India. These are commonly known as carp minnows. They are hard and have no food value. They are useful as larvicidal forms.

Resbora daniconius : It is a small sized fish (3- 13 cm in length) and found throught the fresh water of India. It is regarded as an effective larvicidal fish.

Anabas testudineus : It is a large species measuring upto 30 cm in length. It is a very hardy fish and may grow well in oxygen depleted waters of ponds, sewage and canals. It is reported to be a useful larvicidal fish and is of great economical value also.

Etroplus : It is represented by two species in India. It lives in brackish water. The yong ones feed on insect larvae but adults feed upon algae.

Mugil : It is commonly known as mullets. Some species live in estuarine and fresh waters. The young ones feed on mosquito larvae.

Badis : It is beautifully colourd fish measuring 5-8 cm in length. They are reported to feed on insect larvae.

Oxygaster : It is commonly known as chilwa. There are several species of *Oxygaster.* The smaller varieties feed on mosquito larvae. They are delicate fishes and the large species are used as food.

Besides above, *Laubuca, Ambasis, Therapon jarbus* etc. are also regarded useful in the control of mosquitoes.

Classification based on larvicidal activity

On the basis of mosquitocidal activity, the larvicidal fishes are classified into following catagories by *Hora* and *Mukerjee* (1953). These are -

1. Typical surface feeders

Fishes such as *Aplochielus* and *Gambusia* fall in this catagory. These fishes satisfly all the requirements of larvicidal fishes, hence are highly efficient for antimalania work. *Oryzia, Aphanius,* and *Horaichthys* are also surface feeders but are relatively less efficient due to their mode of life.

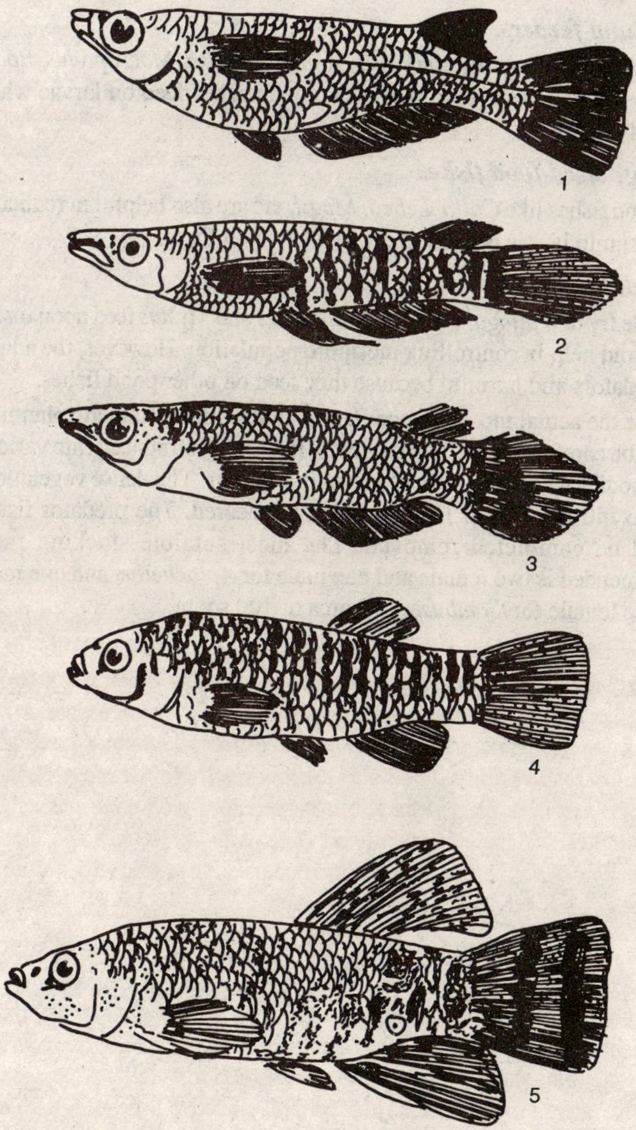

Fig. 15.5. Larvivorous fishes. 1. *Oryzias metastigma*, 2. *Aplocheilus lineatus*, 3. *Aplocheilus panchas*, 4. *Aphanius dispar* (female), 5. *Aphanius disper* (male).

2. *Sub-surface feeders*

Dano, Rasbora, Esomus and *Carassius* are easy to raise but are larvicidal to a considerable extent.

3. *Column feeders*

Fishes like *Punctius, Amhassis, Anabas, Colisa, Macropodus, Badis* and *Therapon* are included in this catagony. They feed on larvae when they get chance.

4. *Large sized food fishes*

Food fishes like *Catla, Lebeo, Mugil,* etc are also helpful in reducing the mosquito larvae because their fry feed upon them.

5. *Predator fishes*

The fry of *Wallago, Channa, Notopterus* and *Mystus* feed upon insect larvae and help in controlling mosquito population. However, the adults are predators and harmful because they feed on other pond fishes.

For the actual mosquito population control the systametic planning should be carefully done. The larvicidal fishes are introduced into various water bodies where breeding of mosquitoes occur. The dense vegetation, prior to introduction of fishes, should be cleared. The predator fishes should be completed removed. The most sutaible stocking ratio recommended is two female and one male for *Aplocheilus* and one male and one female for *Gambusia* in a area of 100 sq. m.

16

Migration in Fishes

A man migrates when he quits one country, and immigrates when he enters another to settle : these are one way trips and the traveller need but a single ticket. The word migration is often used in the sense of moving from one place to another. But for animals which live for a year or more, migration has a special meaning. There is a comming and going with the seasons- even if not always on an annual basis- and the traveller needs a return ticket.

There are about 8000 species of fish which live in freshwater and a further 12000 which live in the sea, and there are about 120 diadromous species which move regularly between the two environments. Fishes are not distributed evenly throughout the world. About 100 species are found with in the Antartic convergence which suggests that there may not be more than 200 or 300 species in polar latitudes. There are many more species in temperate waters but the majority, probably at least 70% of the total, are found within tropics where over 500 species may be found on a single reef.

Many fishes are restricted in that they are limited to one geographical area with in which they make wintering, spawning and feeding migration. These movements are relatively short. But some species, perhaps no more than a few hundred, include stocks which migrate between widely seperated areas and here the distances involved may be substantial. The distinction is between distances of 20 to 30 km and those of several hundreds or even a few thousand kilometers.

The term "migration" is frequently employed to express three different kinds of movements. In the early stages of life the eggs and fry may be carried by the currents some distance from the places where they were born. This helpless drift is at first involuntary, but later the young fry are quiteable to make their way in and out of currents or remain stationary. Then the young fishes and the adults spend most of their time seeking food, they scatter about within their area; but good feeding grounds or incoming currents such in plankton food bring many together in a concentration shoal which persists for some times. These may be called *voluntary movements* on the part of the fish.

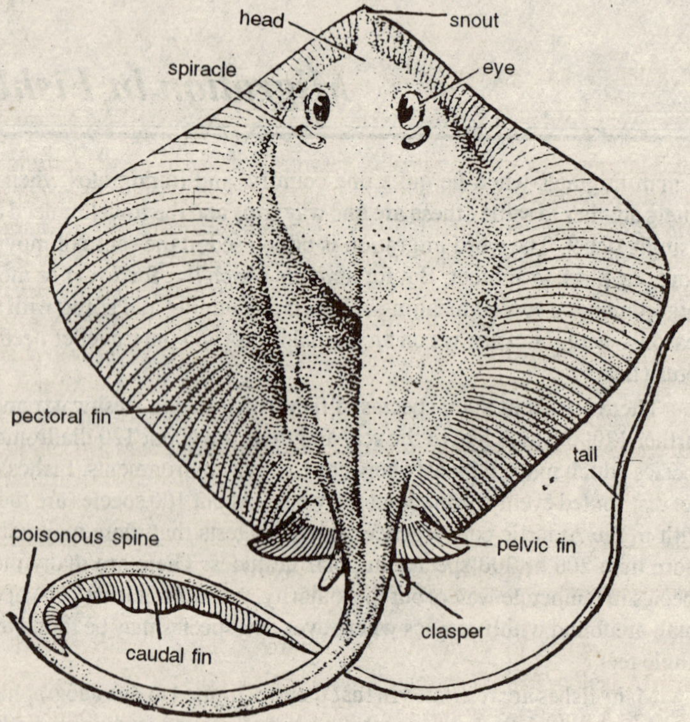

Fig. 16.1. *Tragon.*

Not all fishes are migratory. In fact a majority of them ristrict their movements to small territorial limits and do not go out of these during any part of their lives. Few species, however, travel over long distances for feeding and spawning either in the same type water to which they belong or in search of water naturally different than their own, e.g from fresh water to sea water or vice versa.

Rightly considered, the true migration is something very different from the drift and wanderings. Here the fish is urged to leave its ordinary hunting grounds and even to stop feeding, in order to undertake sometimes long and perlous journeys. The call of the blood, a deeper necessity common to all living things, takes possession of the fish, and the thousand

and millions in like case for gather year by year almost at same place and time a circumstance of great importance both in the past and in the present.

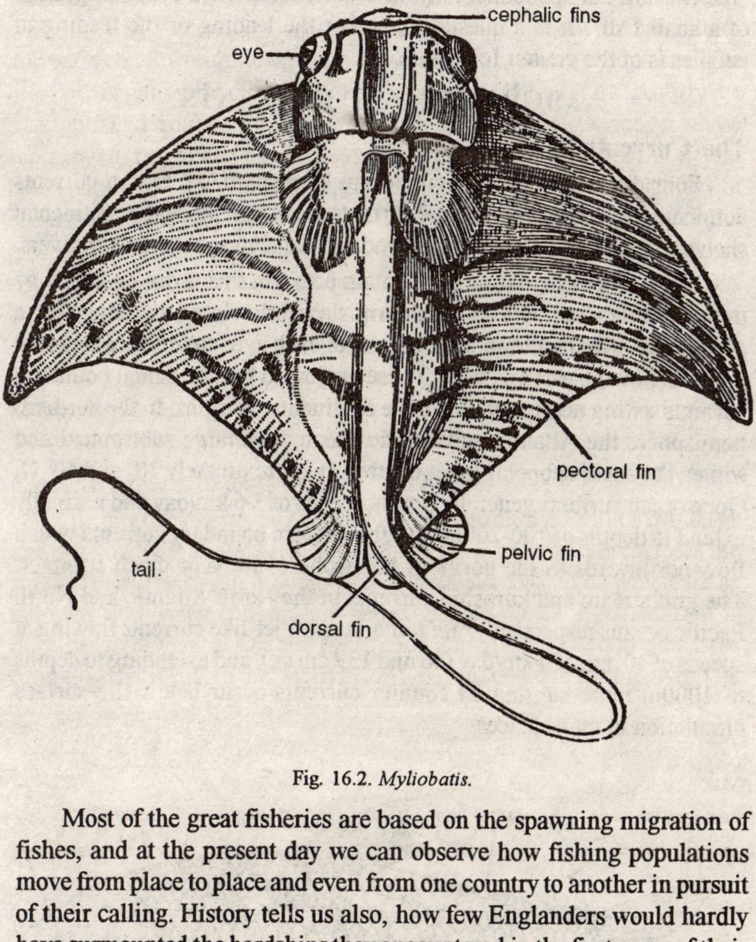

Fig. 16.2. *Myliobatis.*

Most of the great fisheries are based on the spawning migration of fishes, and at the present day we can observe how fishing populations move from place to place and even from one country to another in pursuit of their calling. History tells us also, how few Englanders would hardly have surmounted the hardships they encountered in the first years of their endeavours to settle in America but for the rich fusheries of the coast, and again, how the Dutch fishing population spread from one country to another in the seventeenth century, and how England and Holland went to war on a small matter of the fishing industry. In ancient times also the Greek and Phoenician colonists chose their settlements where the fishing was of importance, and the nature of the fishing is sufficiently indicate by the representation of the Tunny on the old medals of carthage and cadiz. These under currents in the life of nations are not apparent in the surface histories yet *Edward Forbes* was justified in his quaint comment, "the existence of Alexender may have been dertemined by the migration of a shall fish". It is a question whether the leading or the feeding of peoples is of the greater importance.

MIGRATION AND CURRENTS

The Currents

Four distinct currents regimes can be recognized. Ocean currents dominate the high sea, while tidal currents are important on the continental shelves. In the estuaries the discharged supplements the ebb tide, in rivers.

The surface circulation of the main ocean basins is driven argely by the prevaling winds. Its basic pattern is a nearly closed system called a *gyre*. The northern and southern limits of each gyre are marked by zonal currents flowing east and west. These are joined by meridonial boundary currents setting nearly parallel to the continental margins. In the northern hemisphere the Atlantic and Pacific ocean have large subtropical and somewhat small subpolar gyres centred at approximately 30^0 and 50^0 N. Open ocean currents generally flow at speeds of 3-6 km/day and normally extend to depths of 100-200m. But the western boundary currents which flow northwards in the northern hemisphere, are very much stronger. The gulf stream and kuroshio currents in the North Atlantic and North Pacific oceans respectively, are rather narrow jet-like currents flowing at speeds of 40 and 120 km/day (46 and 139 cm s^{-1}), and extending to depths of 1000m more substantial counter currents occur below the surface circulation in many places.

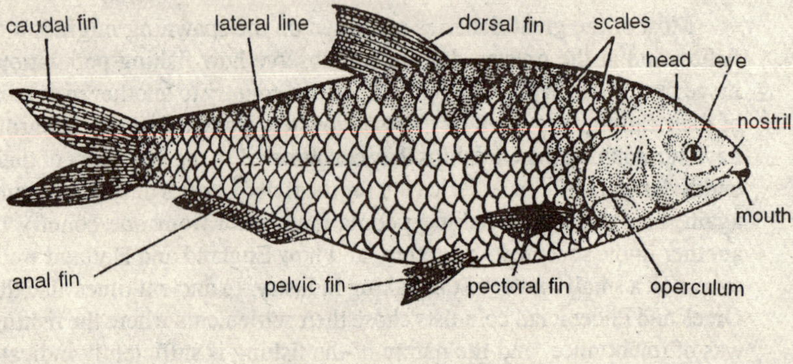

Fig. 16.3. *HIlsa ilisha.*

Tidal currents, which are generated by the combined gravitational effects of sum and moon, are present in both oceanic and shelf waters. But their effects are much more pronounced on the continental shelves. Near coasts the rotatory tides of the open sea are replaced by familiar revesing tides of the seaside. Flood tides, where the water flows towards coasts are saperated by periods of slack water from ebbtides, where water flows away from the coast. Three types of tides can be recognized : diurnal tides with one high and one low water per tidal day of 24 h 50 min. Semidiurnal tides with two high and one low water per tidal day and a tidal period of 12h 25 min and mixed tides with appreciable differences in height between successive high and low tides and unequal tidal periods. Semidiurnal tides are typical of the Atlantic coasts of Europe and North America, mixed tides of the Pacific coast of North America. The daily tidal range is greatest during spring tides which occur at times of new and full moon and least during neaptides which occur near the first and third luman quarters. The speeds of the tidal currents vary correspondingly.

An estuary is a semi-enclosed tidal basin in which the circulation is a mixing system of fresh water and salt water. In a simple salt wedge estuary, where the river flow is dominant, there is a vertical stratification. A wedge of sea water extending upstream on the bottom is seperated by a sharp discontinuity form fresh water flowing sea wards above. Where the river flow is less dominant there is still a two layer circulation but entestainment and vertical mixing break down the interface and produce a salinity gradient which may be localized in the halocline or which may extend throughout the water column. Where the tidal current is dominant the intensity of vertical mixing is such that both the salinity gradient and the vertical stratification break down. The seaward and landward net flows are then seperated horizontally rather than vertically. Seasonal fluctuation in river discharge very the degree of tidal influence.

Reactions to the current

Flat fish lying on the sea bed can detect the currents as a result of direct mechanical stimulation. But a fish not in contact with the bottom requires a visual reference point and must be within a few meters of the river bank or sea bed to detect its displacement. Fish can detect currents speeds of 1-10 cms^{-1}.

There is no evidence that fish out of sight of the bottom can detect the currents. Westerberg's (1979) observations are inconclusive but fish in midwater is the sea might be able to orientate visually at the rheocline of particles are visible in the water. Alternatively they might be able to detect the electric field generated by the movement of water through the vertical components of the earth magnetic field. Skates and rays and salmon and eels can detect potential grational of 0.01-0.1 μcm[1]; gradients of 0.05-0.5 μcm^{-1} commonly occur in the ocean.

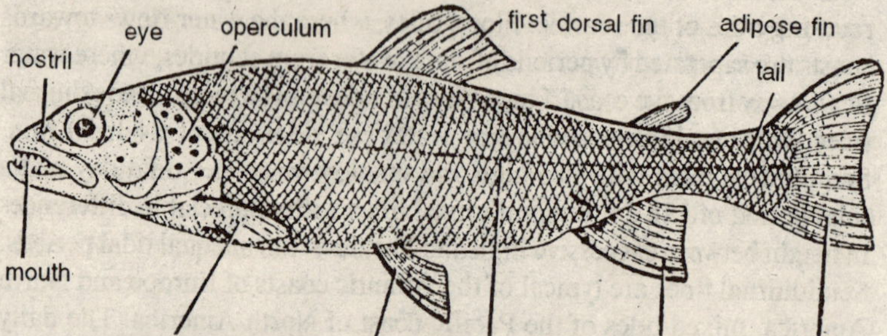

Fig. 16.4. *Salmo.*

The basic reaction of fish to currents is rheotaxis. Most fish turn to head into the current and maintain station relative to a fixed reference point. As the speed the cirrent increases the fish swims faster, a kinesis is superimposed on the taxis. But this behaviour must be modified for migration and at different stages of their life history diadromous fish move both upstream and downstream. Loss of visual reference points at night may account for the downstream displacement of salmonid fry. But the seaward movements of smolts and adult eels are apparently active oriented movements; the fish head downstream and swim with current.

There are some evidences that diadromous fish show behavioural changes at the interface between the currents regimes. A delay at this freshwater saltwater boundary is commonly in both adult and juvenile salmonids. Silver eels change from negative to positive rheotaxis on exposure to saltwater. Newly metamorphosed elveis at first drift passively with the tide in coastal waters, progressing only with the residual current subsequently they switch successively to tidal transport and active upstream movement. The second change occurs in the River Elbe at least, at the limit of the tidal incursion.

Oceanic migrations are generally consistent with movements with, rather than against, the current. Because they are not well described in relation to oceanographic features, they can be accounted for by a number of different models. These variously invoke passive drift, active dounskeam migration are some forms of independtent navigation. Passive drift in the North equitorial current and counter current can, for example, account for the movements of young skipjack tuna, *Katsuwonus pelamis*, in the Pacific ocean and jet stream transport can explain those speeds which are significatly faster than average. Two blue fin tuna, *Thunna* thymmus, which travelled at least 4832 km between the Bahamas and Bergenin Norway at an average speed of 40 km/day probably received substantial assistance (from the Gulf stream and North Atlantic) drift. But in several instance ground speeds estimated from conventional tagging

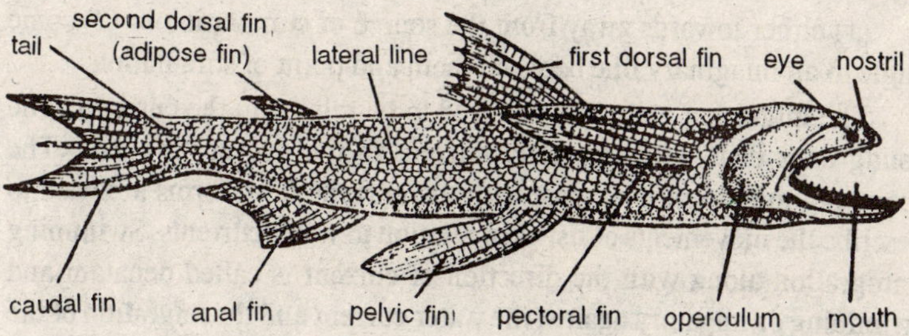

Fig. 16.5. *Harpodon nehereus.*

experiments suggest that oceanic migrations can not be accounted for by drift alone. Sockeye salmon, *Oncorhynchus nerka*, in the eastern North Pacific ocean travel at speed ranging from 20 to 50 km/day. Two albacore covered 5239 km from the central north Pacific to the coast of Washington State in 110 days at an average speed of 48 km/day. Because surface currents in these areas have much lower average speeds it is argued that these are active oriented movement. *Royce*, Smith and *Hartt* suggest that salmon follow the ocean circulation in the North Pacific but swim dewnstream faster than the current, which they detect by means of electric potentials.

Heape (1931) has defined migration as "a class of movement which impels migrants to return to the region from which they have migrated."

Migration can be said to typed as following :

1. Alimental migration for the need of food and water
2. Gametic migration for reproduction and development
3. Climatic migration for securing sutiable climatic condition
4. Osmoregulatory migration for better watery conditions to live as for osmoregulation is concerned.

A fish can make migratory movements by several methods.

(A) ***By drifting*** The fishes are carried passively, by water currents. This may result in directional movement if the overall water movement is in one direction.

(B) ***Random locomotory movements*** Random movements in any direction may lead to distribution or aggregation. If fishes move uniformly from one point to all directions, it is called dispersal which results in distribution.

(C) ***Oriented swimming movements*** The fishes swim in a particular direction.

(1) either towards away from the source of stimulation or (2) some angle to an imaginary line between them and point of stimulation.

The migration pattern is believed to be related to the currents. The young ones drift with the water current to the nursery grounds. The spawning migration is against with the current. Two terms are used to describe the movements of fishes in relation to water currents. Swimming a migration along with the direction of current is called denatant and swimming or migatory against the water current and the migration of the adult fishes towards the spawning area is called contranatant.

TYPES OF MIGRATORY FISHES

Anadromous

Those fishes which spend a major part of their lives in the sea but mugrate to freshwater during breeding season for spawning. Salmon and Hilsa have been found to travel several/thousand miles in the sea and then several hundred miles inland to reach the spawning grounds.

Movements of adult fish

Adult plaice fitted with 300 KHz transponding acoustic tags and tracked in the open sea by sector scanning. Sonai show a pattern of semi-diurnal vertical movement related to the tides which we have called selective tidal stream transport.

The fish were tracked within an area bounded by latithdes, S1° 40'N and 52° (40"N longitudes 1° 40' E and 2°) 30° E. The tides in this area are east, with speeds up to 1.4 ms^{-1} at spring and highly directional, flowing approximantly north, north east and south, south east. The tide floods on the south going stream and the tidal range is 3m in water of 20-40m depth. The bottom deposits are mainly fine to medium sand with beds of gravel in some places and patches of sand waves in others. The water is generally turbid and visibility is limited to few meters.

The fishes were caught by olter trawl in the silver pits and on the Borkum Ground and were help in laboraloly for about two months prior to release. Twelve fishes were released singly and followed for periods of 25-55 h and distances of 9-61 km. Eight fishes which covered distances greater than 15 km showed a consistent pattern of tidal transport. They made regular vertical migrations coming off the bottom at one slack water and remaining in midwater for the following 5-6h before returning to the bottom of the next slack water. When in midwater the fish moved downstream with the tide for distances of 12 -16km, more or less equal to that of the tidal displacement. Activity was limited when the fishes were on the bottom and some fishes did not move for 2-3h. The regular cycle of movements on and off the bottom at slack water was very striking and this pattern of semi-diurmal vertical migration was characterstic of all fishes that move appreciable distances.

The trach of plaice no. 7 was typical. The fish came up into midwater after high water slack, gained ground tide and returned to the bottom low water slack. The speed of the fish over the ground was 890, 68 and 105 cm s^{-1} for the three northerly tides compared with only 8 and 9 cms^{-1} for the two south going tides.

During this, and some other tranks, water current measurements were made from a second research ship or with moorved current meters, from these measurements it was possible to estimate the speed of the fish through the water. It rarely exceeded one body length per second and, while fish no 7 appeared to swim downstream with the tide three other fish swarm through the water heading consistently southeast as they were being carried northwards over the ground.The orientation was maintained for periods up to one hour and may well have been more determined with reference to sandwanes before the fish left the bottom.

The spawning migrations

It plaice used selective tidal stream transport when on migration, ripening fish moving towards their spawning grounds in the southern. Bright during November and December should be in midwater on south going tides, where as spent fish returning to their feeding areas in February and March should be in mid water on north going tides. This prediction has been tested by midwater travling experiments along the line of the western migration route.

The experiments were carried out with large and midwater trawls. pairea tows, usually of 3h duration, were made on successive tides and the gear was shot so as to start fishing at the required depth one and a half hours before the expected time of maximum tidal speed. Each haul was made downtide along the tidal axis and so for as was possible, the research vessel covered the same track over the ground on each two of a pair.

The work was carried out on the western side of the southern. Bright and most places were caught in the line of the small stream between latitudes 52° N and 52° 40' N, longitudes 2°00' E and 2° 30' E. At the centre of this area the directions of the north going and south going tides are 020° T and 200° T, respectively, peak speeds of 70 and 120 cm s^{-1} are reached at reef and spring tides.

Fifty paired tows were made in November and December and 22 in February giving 39 and 21 pairs, respectively, in which place were caught in one or both tows. More plaice were caught by night than by day but this diurnal effect was removed by restricting the comparison to the 17 autumn and 8 'winter' pairs in which both tows were completed during the day. During the prespawning migration singnificantly more fishes were caught on the north going tide.

In the autumn most of the female plaice were still maturing and none was yet ripe, whilst half the males were already running ripe and extruding sperms. No spent fishes were caught in autumn but three quaters of the female caught in February had completed spawning virtually none of these had been feeding.

Diurnal vertical movements on spawning and feeding grounds

Plaice kept in aquarea under natural illumination show a dear rhythm of diurnal vertical migration on, feeding on the bottom during the day and swimming freely off the bottom at night. During most of the year catches of plaice in bottom travls in the southern right are lower by night than by day with an average day. Night catch ratio 1:3:1:0. But on spawning grounds, in January and February this ratio is revesed plaice are quite commonly caught well above the bottom in midwater travels and it is generally asumed that the reduction in catch at night is partly on mainly the result of a diurnal vertical migration, where by at night a propotion of the population is on longer accessible to gear on the seabed. The reversal of the catch ratio during the spawning season may perhaps be attributed to an alteration or even a reversal of this behaviour.

Movements of juvenile fish

Selective tidal stream transport is probably an innate pattern of behaviour in the plaice, occuring as it does with other rheotactic behaviour at metamorphosis immediately after the larvae have gone to the sea bed. It has been demonstrated for larvae entering the Wadden sea in March and April through the marsdiep inlet between Texel and Den Helder.

Catches of larvae in an Issacs-kidd midwater travel were consistently higher on the flood than on the ebb tide and the catch ratio increased progressively with the age of the fish less than 20% of fully metamorphosed stage 5 larvae were found in midwater on the ebb tide compared with over 65% of the early stage 4 larvae, which are only just begining to metamorphose.

After spending their first summer in the Wadden sea the o-group plaice move out to sea again, if the autumn to overwinter offshore. Some return again as I group fish in the following February and March. These fishes also use selective tidal steam transport, moving landwards through the marsdeip and vliestroom inlets on the flood tide.

As both O and I group fishes the plaice in the Wadden sea show tidal feeding migration, collecting in the creeks at low tide but moving up onto the mudflat as they are covered by the resing tide. Similar tidal migrations occur on Sandy beaches.

MIGRATION OF DIFFERENT FISHES

Aristotle and the ancient writens presumably echoed the common belief of those days in saying, that the tunna migrated from the Atlantic along the north coast of Africa to Egypt and Palestine and thence into the Black sea. It was supposed to spawn in the sea of Azov, thence returning along the northern coasts of the mediterean out into the Atlantic. Such wide reaching ideal with regard to the migrations of fishes as well, it is their trebute to the mystery of their calling. But the scientific investigators of the past century have greatly reduced the limits of the migrations except in one remartable instance to be noted later.

The presence of the young in all stages and the appearance of the adults in all parts of the Mediterrean at practically the same time (summer), are among the phenomenon which is to beleive that the Tunny lives for most of the year in the deep water off the coasts. As the spawning time approaches, the fish first of all makes a vertical migration from the depths of the surface. It is said that new arrivals in the surface water can be distinguished from the earlier by their dirty colours. Then they move towards the shallower water to spawn. It is the latter, the horizontal movement, which has given rise to the idea of immigration from the Atlantic and which accounts for the elaborate preparations made by the fisherman for the reception of the tunny.

These are not spawners, they are the larger fish which, after spawning in the summer on the Spanish or Portugese coasts, follow the sardine and mackeral shoals northwards and later feed on the autumn Herring of the

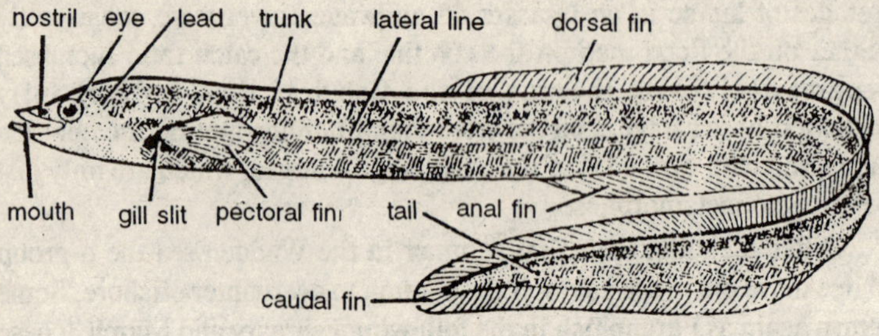

Fig. 16.6. *Anguilla.*

North sea. When the travel with its load of Herring is hauled up the Tunny follows and snaps at the protruding neads and fails. Possibly it was just as common in earlier years, there is a record by *Schonevelde* that it was fairly common near Echernforde in the western Baltic as for back as the beginning of the seventeenth centuary, also in connection with the presence of mackerel. The propoise may often have been blamed for its depredations, but until the travel was used for the capture of Herrings in large quantties, its occurance was regarded as a rarity.

A curious thing about the Herring not known in any other species, is its ability to spawn at all seasons of the year and in fresh as well as salt waters thus under the most diverse conditions as to temperature and salinity. Its variablity in this respect suggests that from the beginning it has been a shallow water or brackish water form, it can readily adapt itself to any conditions from 0 down to 100 fathoms. There are winter Herrings which spawn in the firth and forth and in the west; spring Herring off the north east coast, autumn Herring spawning in the sourthern north sea, and so on. But the Herring keeps to the temperate zone, in the tropical and subtropical regions the variability has changed it into other species. Another interesting problem in distribution is furnished by its relative, the Anchovy. This is a southern, subtropical form, with its main spawning places in the Mediterrean and adjacent water yet- a Northern branch appears regurlarly year by year, and that in great quantities, in the zuidersee and neighbouring brackish waters. In southern regions the species spawn in water of comparatively high temperature and salinity, here the water is brackish and cold. There is a wide gape between Spain and the Zuiderseae, few specimens have been taken in the channel or Bay of Biscaly Arewe to suppose that it migrates this distance yearly, or that it remains somewhere in the channel, where it eludes capture. In either case the origin of this branch remains a mystery, possibly it represents the original form of Anchovy, as we can more readily understand a migration to the pleasanter conditions of southern waters than the reverse. On the other hand, this particular branch may have suffered greatly from its enemies, mackeral.

Tunny and others and found a safe retreat from their puisuit in the Zudasee. But no definite soultion can be given to such a problem; of greater interest would be the investigation of the structural and developmental differences between forms the same species growing up under such wide differences. The principal thing about the species so for mentioned is, that the migrations are from deeper to shallower, water in the sea. Some clupeids like the sheds, go further and second rivers to spawn in fresh water. Some are permanent inhabitants of the inland seas.

The same phenomena are found among the nearly allied Salmonoids. Whilst some forms, like *Argentina*, spawn in deep water in the open sea, and others, like Mallotus (the Capelan), come close inshore to spawn thus constituting a most important source of food, for example, for the Eskimos of Greenland, others like the Smelt enter brackish and fresh water, and the principal species of the group ascend rivers right to their sources, whilst whole families belong only to the fresh water.

The *Salmon,* it is now believed, does not move far from the mouth of the river to which it belongs; yet, like the *Anchovy*, it is rarely found in the sea except where it assembles to feed, as in the straits of the Pacific and in the Southern or South-eastern Baltic. The latter case is of near interest; the *Salmon* are taken on hoods, but some escape and are caught later, with the hoods still attched, in the lakes of Finland. The *Salmon* is thus able to spend its whole life in fresh or nearly fresh water. In the Rhine again, when it was a clean river, the winter *Salmon* used to remain nearly a whole year in the middle reaches below Basel before proceeding to spawn.

It sounds almost incredible, therefore, that the *Salmon* does not feed after it has entered fresh water. Yet the evidence from all countries supports this view, and further, the digestive organs undergo a certain amount of degeneration. How then can we explain that the *Salmon* rises to the fly and has even been taken with a trout in its stomach? For one thing, it is certain that the *Salmon* of different regions, even different rivers, vary greatly in their powers and habits; for another, we hae only to take a wider perspective to understand what happens. In the larger Norwegian rivers the *Salmon* rise to the fly in the lower reaches, but provide no sport in the upper. In the still longer American waters the *Salmon* provide no sport at all. They feed, therefore, or endeavour to feed only when fresh from the sea.

All the time they are in fresh water, and it may be a whole year, they live on the materials stored up in their body. This has been considered a wonderful provision or adaptaion of nature. Regarded from the Salmon's point of view, it seems more like a tragedy; the *Salmon* wishes to feed but cannot ! Certainly, its body must be an accumulator of energy of very good quality and enormous power. The fish ascends rivers against strongly running waters, jumps numerous falls (probably not more than10 ft. at a single leap however) ; if a male it has to fight other males and keep off enemies, then helps to prepare a bed in which the female lays its eggs. It is not to be wondered at, when the spawning is over, that the fish is completely exhausted and allows itselt to drift tail-first down the stream.

The Atlantic salmons spawn mainly during November and December. They ascend the rivers as the breeding period approaches, travelling several thousand miles in sea and then inland. On entering freshwater they stop feeding and lose weight. Their bright silvery colour changes to a dull reddish-brown shade and the skin becomes thick and spongy. The body of the male *Salmon* becomes spotted with red, orange and large black spots. After spawning, the return journey to the sea begins but few males survive to breed second time. However, many females are able to reach the sea and start feeding. They soon recover their normal condition and silvery colour. The *Salmon* does not usually spawn more than three times in its life-span of eight or nine years. From the time it reaches freshwater till spawning, the *Salmon* lives on its reserve food accumulated in its beop in the form of fat.

The king *Salmon* of the Pacific coast of North America travels about 2250 miles from the sea in order to reach the rivers to spawn. However, all of them die after spawning. No male or female survives to undertake the return journey.

Studies have shown that young *Salmon* hatched in freshwater cannot return to the sea until their salt-secreting cells have developed. Studies indicate that migratory behaviour in both adults and the juveniles is not due to any special hormone regulation other then that which influences their general metabolism. Changes in their body states and behaviour may expose them to different environmental conditions to which they react in specific ways, such as travelling up freshwater streams by adults and downstream by juveniles.

Several theories have been proposed to account for the unique way salmons are able to return to the very place where they were hatched.

Some authorities believe that the fish have an extremely sensitive chemical sense that enable them to perform this feat. Another theory links the accelerated metabolism at spawning time with the need for more oxygen which increases as they ascend the hard waters of a stream. Still another idea stresses the importance of a carbon dioxide gradient in the water as being the determining factor. Recent researches strongly favour the theory that sense of smell in fishes is the determining factor in locating home grounds, for different streams have different odours.

The *Hilsa* of the Indian Ocean also migrates to rivers for spawning. This species is known to ascend up the Ganges and go upto Varanasi and Allababad. During the monsoon months *Hilsa* also migrates to many rivers of southern India. However, the details of its migratory habits are not known.

On the whole, it may be said that the run of *Salmon* up the rivers is connected with spawning, yet some important exceptions have to be made. Many specimens will go some distance up a river only to turn back (and this may happen several times), or even remain months in definitely fresh water before proceeding to spawn. There may be a simple explanation of these eccentricities : it must be remembered that the *Salmon* is not far from being a freshwater fish. According to *Paton* and *Newbigin* (1900) the factor determining migration from sea to river is not the "*nisus generativus*, but the state of nutrition". The spawning comes later.

With the *Salmon* we may finish this account of the one stream of migratory fishes, those which journey from deeper to shallower and even fresh water in order to fulfil their life purpose. This is by far the most common form of migration ; a very large number of marine fishes, even the Sharks, Skates, and Rays. follow in the main the same course. As suggested by *Meek*, the term "anadremous," formerly used only for those entering fresh water, may well be applied to all those fishes which follow the same general rule in their migration, namely, an ascent from deeper to shallower waters. To a very large extent, this at once indicates the nature of the problem to be solved and even suggests the solution.

But there are two groups of apparent exceptions. In the one the fishes exhibit but little tendency to undertake any spawning migrations. They spawn, as it were, where they find themselves. This applies naturally to the stationary fishes, though even among these a certain amount of migration can be noted. Thus, among the Blennies the Butter-fish or Gunnel has a short, anadromous migration of a few miles, whilst the *Angler* or Frog-fish shows a longer migration in the reverse direction. Among the Gadoids, again, most species prefer the deep water for spawning ; but the Cod. if it has any preference, comes from the deep water to shallower banks.

These groups, in which the migratory movements are of little account or in which diffrent tendencies are shown by different species, may be called neutral for the moment. The probability is, indeed, that they are not so much affected by the spawning process. The seeming migrations may be either a return to the natural home places, following upon the drift of the young away from these or they may be mixed up with the movements in search of food. The Cod, for example, is a rover with a circumpolar distribution like the Herring, and it shows a decided preference for the lattar as food. It pursues the latter at all seasons of the year and at the proper time it is able to spawn whenever it goes-within certain limits, of course, but they are very wide and correspond closely to those of the Herring.

From the biological standpoint these neutral groups are of great importance in any theory regarding the causes of spawning migrations. We may take it, that the species composing them are so well-adapted to their surroundings that they are not urged to make any great change when the need of spawn comes upon them. Viewing their life-history as a whole, they show a better balance between the inner and outer conditions.

The second exception is represented by the Eel group. With the Common Flounder the Eel constitutes the old group of catadromous fishes, those which migrate from the fresh water to spawn in the sea. The Flounder is certainly catadromous like the Plaice seeks water of greater density than its ordinary feeding grounds for the purpose of spawning ; but other Flat- fishes like the *Soles* show the reverse procedure, and it seems better to regard the group as neutral. There is in any case no comparison with the lift-history and migrations of the Eel.

The Eel seems to belong to quite different world from other fishes : it would be difficult to imagine a greater contrast. Whilst other fishes are content with a reasonable area of distribution, thus admitting the limits of the piscine nature, the Eel ranges between the most extreme conditions on the globe, from mountain stream to the depths of the Atlantic. It passes through two transformations in the course of its life, when it enters and when it leaves fresh water, other fishes have only one or none at all. The Eel takes some ten to fourteen years (in fresh water, plus three more in salt water) to become mature, other fishes everywhere only one to three.

Other Eels, Conger, Muraena, etc, do not show such a great contrast, so far as we yet know their life-histories. All seem to be catadromous fishes, migrating out from shallow to deeper water, but only the *Anguilla* genus enters fresh water.

Eels have two distinct phases in their life-history, each morphologically different from the other. Yellow eels represent the feeding and spawning phase while silver eels represent breeding phase. Yellow eels are found both in salt and freshwater and vary in length from a few inches to five or six feet. During autumn a number of yellow eels become silvery and prepare to leave for spawning grounds. They stop feeding, their eyes become large, lips thinner, and the snouts sharper. Their colour becomes silvery along the sides with a blackish back. The reproductive organs develop while the alimentary canal shrinks. The eels migrate down the rivers to reach the sea. They travel about three to four thousand miles in the Atlantic to reach their breeding grounds in the western Atlantic, south of Bermuda. It is believed that the eels spawn at a depth of about 400 metres below the surface and the parents die soon after.

The young larvae that hatch out are called leptocephati. They are flat, leaf-like, tiny creatures provided with long needle-shaped teeth for feeding. They begin their homeward journey in the easterly direction. They grow rapidly during the first few months and when reach the coast of western Europe they are about three inches long and two years old. At this point they undergo metamorphosis and stop feeding. The needle-like teeth are lost and the transparent body assumes a cylindrical shape. They are now called elvers, or glass eels. The elvers, when about three years old, start ascending the rivers in large numbers. They are able to cross all obstacles and reach their suitable resting places where they feed and grow for some years to develop into yellow eels. Finally on attaining maturity, they change into silver eels and start moving towards their breeding site in the sea like their parents.

The American species, *Anguilla rostrata* also has similar breeding habits. In the western Atlantic larvae of both species are found living together. The interesting and mysterous aspects of this strage migratory phenomenon are, how the tiny elvers whithout any guidance are able ot find their way acress the sea, and why the two species which practically hatch in the same region do not get mixed. The explanation lies in the fact that the American eel grows more rapidly (only in one year as against three years in the case of the European species). Thus if the larva of the European eel travels westward instead of an easterly direction, it will reach the coast of America long before it is ready to change into an elver. And conversely, if the larva of the American eel migrates in an easterly direction it will undergo metamorphosis in the middle of the Atlantic. In other words, the larval life in each is geared to the distance it has to travel.

A recent theory, however, suggests that both American and European eels belong to the same species and that the adult European forms die before reaching the spawning grounds. All eggs laid by American eels but, when hatched, the larvae in the northern part of the spawning grounds are carried away by currents to Europe. Those in the southern part are carried towards the American coast.

By a series of brilliant observations and deductions *Johs Schmidt* (1921) has at length traced the Eel to its spawning places in an indirect yet quite convincing manner. The post-larval Eels (*Leptocephali*) were found in ever increasing quantities the further he proceeded westwards, but they became gradually smaller. Successive lines or zones could thus be marked off, according to the size, until at length he found the smallest larva with the yolk-sac still attached and even the eggs out over the deepest parts of the Atlantic, near the West Indies and nearar the American than the European coasts. These were taken in the middle and upper layers, and there is therefore some doubt whether the adults actually spawn at the bottom. The depths on the "spawning grounds" were about 3000 fathoms.

Two more of Schmidt's discoveries are of interest. At the same place where he found the larvae of the European Eel he also took those of the American Eel. The spawning grounds of two forms. living thousands of. miles apart, are practically the same, and the young of both species were occasionally taken in the same haul of the net.

The second discovery is even wore extraordinary. The two streams of post larvae spread out and move northwards towards the Bermudas ; there they separate, one going east to Europe, the other west to America. The latter form spends only one year in the Atlantic and is then ready to enter fresh water ; the European form takes three years to cross the Atlantic before it is ready to change its constitution and enter the rivers. How these tiny fishes can settle which direction to take, even allowing for a difference in constitution, is a profound mystery. In illustration of the immense quantities of these young *Leptocephali* it may be mentioned that *Johs. Schmidt* took as many as eight hundred specimens in one haul out in mid-Atlantic.

FACTORS AFFECTING MIGRATION

Many efforts have been made to connect the migrations of fishes with the external conditions, temperature, salinity, light, etc., and it is probable that each has some influence, but the problem has another side. In the first place, it is essential to distinguish the food concentrations and wanderings from the spawaing migrations, a matter not so easy as at first appeared. It would seem, indeed, that a feeding concentration, in many cases if not at all, precedes the actual spawning, and it may be that the storing of foodstuffs leads directly to the internal changes in metabolism which give the stimulus to change of grounds and conditions (*Paton*, 1900).

Physiologically the feeding and the migratory impulses are very different. In the one the fish is seeking to obtain something, in the other to get rid of something. The latter may be called a morbid, even pathological condition, affecting every part and sufficiently powerful to make the fish ignore terrible injuries and even to over-come the impulse to feed. The digestive organs partially or wholly degenerate, and the fish is then urged to seek relief by unusual or extraordinary means, that is, to find if possible external conditions in better agreement with its changing or changed constitution.

One cannot say definitely what happens in any particular case, owing mainly to the protean nature of fishes. A dog-fish, for example, is more in harmony with the surrounding medium than a Salmon, and a Stickleback will survive undcr conditions that kill. most fishes. These phenomena depend apparently on their ability to alter the blood-pressure, or concentration of salts, as they pass from one set of conditions to another. And this ability, again, depends upon how far the gill-membrane and the skin permit of the passage of water outward and inwards, a question that has not yet been sufficiently investigated.

It is possible, therefore, that the internal stimulus to migration comes from some internal change in the blood ; probably in the blood-pressure. This is always much less than the pressure of the surrounding water, but it varies. In the Salmon, *Greene* found that the concentration of salts in the blood was far below that of the salt water in which the fishes were feeding voraciously, and gradually altered as the fish entered fresh water. It may be that a slight reduction in the pressure within the blood is a source of discomfort to the fish and urges it to enter the rivers. In the case of Eel, on the the other hand, there would seem to be a great increase of the internal pressure accompanied by an increase in the amount of carbonic acid at the time when the fishes are beginning to migrate from fresh to salt water. If they are detained on their journey in any way, the body thickens and swells and even increases greatly in weight though the fish are apparently not feeding. *Peterson* has shown that the digestive tract is in process of degeneration at this time and even the air-bladder decreases in size. The average Silver Eel in its breeding dress is about 70 cm. (27 to

28 ins.) in length and weighs about $1^1/_2$ lb. The large specimens referred to were from 110 cm. and more, and weighed from 4 to over 8 lb. Most remarkable of all is the great increase in the size of the eyes. These become more than four times their former size and weigh twice as much.

Without being able to say positively that such is the case, one may conclude that these phenomena point to a great increase in the internal concentration of salts and absorption of water. And this increased pressure witliin seems to be the cause of the Eel leaving the rivers to seek the greater pressures of the salt water. When the impulse to migrate comes upon them, the Eels are restless and ill at ease, moving up and down for a time in a manner quite unlike their usual sluggishness, then definitely turning down the rivers. At every stage of their descent they come under increasing pressures without. And it may be recalled here, that the Eels are well-known to be extremely sensitive to the differences of pressure produced by thundery weather.

In the case of marine fishes like the Plaice or Cod, we know that when the spawning fish are brought up in the trawl, the spawn continues to run even when the fishes are still alive. This indicates that the internal pressure persists for some time at the level of the deep water whence the fishes came. Since the low pressure in shore would help in getting rid of the spawn, the migration of the Plaice from shallow to deep water would be unnecessary for that purpose. Hence we may conclude, that an increase of pressure within has occurred with the maturation of the sexual organs, and the resulting discomfort has urged the fish to seek deeper water and greater pressures. Similarly in the case of other fishes. Each species is urged to seek the external conditions, whether of lower or greater pressure, which corresponds best to the change in its own constitution.

17

Parental care in Fishes

The reproductive behaviour in most animals is cyclic and more or less periodic. The reproduction in some fishes occurs only once in its life time. In others it may occur once into 2 to 5 years or in once in 10-15 years. Most of the fishes have an early cycle of reproduction that remains continuous till death. There are some fishes which breed more than once in a year. In all these cases the objective aim of breeding is brought about is union of sperms and eggs. After the egg is fertilized other adaptations then leave to survival of fertilized egg.

Usually the ova and sperms are laid simultaneously to facilitate their union. In open water millions of eggs are laid at one time but all of them do not mature at a time. Some mature in each batch, so that at the final stage of rapid growth of eggs, the female is not required to phase enlargement of belly to extremes due to enlarge sex products. If all the eggs ripe at simultaneously they would occupy a space much larger than the body of the fish.

Spawning

Most fishes have definite season for spawning. Fishes inhibiting warm waters are summer spawners and those inhibiting cold waters are winter spawners. Intermediate types are spring spawners. The spawning season of a particular species or race is meant that period during which most individuals may be found to possess ripe ova or sperms, and this may last only a few days extend over as many weeks or even months.

As the time for spawning approaches the fishes congretes in huge shoals in sutiable localities, and on some ground they may be very closely packed together at this season. Analysis of these shoals generally show the females to be in greater numbers than the males , but this is by no means always the case. Here the individuals of both sexes simply discharge their ova and sperms into the water and the fertilized eggs are subsequently abandoned by the parents and left to the mercy of physical conditions.

The actual rate at which the eggs are extruded varies a good deal in the different species, in some cases all or the large proportions of the ova being ripe for fertilized at more or less the same time, while in others the process is comparatively slow and only a certain number ripen and are extruded at one time.

In the Cod and Plaice, indeed in all our food fishes with the exception of the Herring and Shad (*Alosa*), the eggs are minute and buoyant, and float at or near to the surface of the sea. Under such conditions the haphazard mode of reproduction must inevitably leads to a great waste of sex cells, but this is diminished to a certain extent by the fact that the fishes are closely congregated and that both eggs and sperms will float roughly at the same rate and in the same direction. Drifting about in the sea at the mercy of the wind and currents, many of the eggs serve as food for other fishes. Many are killed by changes of temperature and other physical catastrophes, and many more are cast on the shore by adverse currents. These enormous risks of reduction to which the larval fishes as well as the eggs are subjected, coupled with the difficulty of ensuring that every egg is fertilized after extrusion, are associated with the production of huge numbers of eggs by fishes of this type. A single long mowa, 61 inches long and weighing 54 lbs. was found to have 28,361,000 eggs in the ovaries, a Turbot of 17lbs. weight more than 9,000,000 and a Cod weighing $21^{1}/_{2}$ lbs, 6,652,000; a Flounda, however, produce a mere million ova. On the average, and a sole only 570,000. So great is the destruction of eggs and fry that it has been estimated that in the case of the cod less than one egg in every million liberated ova becomes an adult fish.

Care of Eggs and Sperms

Various methods have been employed for affording care of fertilized eggs and young ones by both sexes. One of the means evolved has been to lay eggs into the safe place. For this purpose the selection of spawning rate is done by fishes before breeding. After the eggs have been laid they require much care till they hatch out. For such care, adaptation in the eggs and the behaviour of the parents have been evolved through which the care of the eggs is exhibited.

Eggs laid without care

Many fishes do not give special care to the eggs. They are adopted themselves to survival. They may float with or without a little filamentous structure through which they may also get entangled. This filamentous structure is transparent and may invisible. In some other cases the eggs may be laid in transparent geletinious mass forming a rope-like structure which is common in perch. In still others the eggs may be laid together inside the depression and further development takes place in nature till they hatch.

Among true fresh water fishes, members of the large and varied carp family produce ova which adhere to weeds, stones and other objects. After spawning no further care of the offsprings is taken by the parents.

PARENTAL CARE

The parental care in fishes can be studied under the following heads.

1. Direct nursuries
2. Indirect nursuries

Direct Nursuries

In a family Solenostomidae the female takes care of the eggs, keeping them in a pouch formed by the pelvic fins. The inside of the chamber being provided with numerous long filamentous processes, which serve to assist in retaining the eggs in position. In the sea. horse, and in the pipe fishes, it is the male who looks after the eggs and young ones. The former being received into the brood pouch beneath the tail, where they remain until hatched. At the breeding season this pouch becomes thickened and well supplied with blood vessels thus being prepared for the reception of the eggs. At the same time the flesh around the opening of the genital ducts in the female becomes somewhat extended to form a genital papilla, which acts as a kind of intromittent organ for the transfer of her ova to the male. The final extrusion of the young fish from the pouch is a much more protracted affair than in the pipe fishes, and may occupy several hours, only five or six individuals being set free at a time. The small aperture of the sac-like pouch makes it impossible for the young to return to its shelter, and it is probable that they are retained for a longer period than in the case of the pipe fishes.

In a special pouch

In pipe fishes (Syngnathidae) the care of the eggs and fry is always under taken by the male fish, who carries them about until hatched, either lodge within a simple groove lined with soft skin in the lower surface of the abdomen, or in a special pouch closed by flaps of skin and situated on the underside of the trunk or tail. In the Florida pipe fish in which the whole process has been observed, mating is preceded by an elaborate courtship. The two fishes swim round in nearly vertical positions but with the head and shoulder region bent forward. They then swim slowly past one another, their bodies come contact, the male bending its body around the female frequently caressing its mate with his snout. Just before the actual transfer of ova takes place, the male becomes voilently excited wringgles its body about in corkscrew fashion, and rubs the belly of the female with its snout. This demonstration is repeated several times, the fishes becoming more and more excited, until finally the transfer occurs, after which the fishes seperate to commence the process again after an interval of a few minutes. During the embrace when the bodies of the two fishes are interwined, the protruding oviduct of the female is rapidly thurst into the small opening at the front end of the males pouch, and the eggs are thus transferred. By a series of contrtions the male then suceeds in moving the eggs to the hinder end of the pouch, and then the whole process is repeated until the pouch is full, after which the male appears to be very exhausted and is quiscent for sometime.

The eggs remain in the pouch until hatched, but even after this event the fry may occupy it for sometime, and, when they are able to swim freely in the sea will still return to its shelter when danger threatens.

On the skin

A cat fish of the rivers of Brazil and Guianas, exhibits a different method of caring for the eggs, but here the female is entirely responsible for their safety. During the breeding season the skin of the lower surface of its body becomes very swollen and tender, assuming a soft spongy condition. As soon as the eggs have been extruded and fertilized, it lies on them, prosses then into this soft tissue, and each egg becomes attached to the skin by a small, stalked cup, remaining thus fixed until hatched. Each cup is well supplied with blood vessels and may assist in nourishing the embryo.

Eggs carried by the bony hooks of males

In the Indo-pacific genus *Kurtus*, the male is provided with a bony hook projecting from the forehead, supported by a special process of the skull. On extrusion the eggs are formed into two bunches by filamentous processes from the egg membranes that one bunch of eggs lies on either side of the male as it swims about in the water.

In mouths

Many of the cichlids projecting their eggs by carrying them in their mouths, thus ensuring their safety and perfect aeration at the same time. This duty is nearly always undertaken by the female, and even after hatching the young fry do not leave the shelter of its mouth. Later, they swim about in the water, keeping always within easy reach of its head, and in danger threaten they return to their refuge with extraordinary rapidily. Most of the sea cat fishes found on the coasts and in the rivers of North and South America have similar habits but here it is always the male who undertakes the care of the eggs. These are produced in small numbers, and are of remarkable size, measuring as much as seventeen or eighteen millimeters in diameter in a species growing to a length of three or four feet. The eggs are carried about until hatched, and the male does not appear to take any food during this period. Mouth, or rather branchial brooding also occurs in the blind cave fishes of the family Amblyopsidae.

Parental care in Bitterling

The bitterling, a small cryprinid, found in the rivers of Europe, takes remarkable precautions to ensure the safety of its offspirng. When the female is ready to spawn the oviduct is drawn out to form a long tube, acting as an ovipositor by means of which the eggs are deposited within the valves of fresh water pond mussels where they are out of reach of enemies. The male fertilizes the eggs after they have been extruded. In this situation they undergo their development, the respiratory current of water produced by the mussel serving to aerate the ova. The fry finally leave their temporary host about a month after the deposition of the eggs. It is interesting to note that the mussels breeding season concides with that of the bitterling, and it is in the habit of throwing off its own embryos into the water, where they become attached to the gills of the bitterling and there undergo their early stages of development.

Parents coiled around eggs

The gunnel or Butterfish an elongate blenny-like fish, some ten inches in length, is in the habit of rolling its mass of spawn into a ball, about the size of a Brazil nut, an operation in which both parents may assist. Afterward one of them remains on guard coiled round the eggs, but it is not certain whether this is the male or the female, or whether both take their turn. Often the mass of eggs, accompained by the parent fish may be found between the valves of empty oyster shells, or in the holes made by boring molluscs in the rocks.

Indirect Nurseries

In Crude nests and Crevices

Apart from the Sticklebacks, nest building is the exception rather than the rule among marine fishes, and any sort of parental care is very rare. Some of the Wrasses (Labridae) are said to construct crude nests of seaweed, shells or stones, an operation in which both sexes take part. Many of the Gobies (Gobiidae), Blennies (Blenniidae), Bull-heads or Sculpins (Cottidae), and Cling-fishes (Gobiesocidae) nearly all of them

Fig. 17.1. Male and female Bitterling (*Rhodeus amarus*) with freshwater pond mussel. The female is about to deposit eggs.

inhabitant of rock-pools between tide-marks or in the case of the freshwater Miller's Thumb (*Cottus gobio*), of swift streams, provide for the safety of their eggs by depositing them in the dead shells of mussels, oysters etc., in crevices in the rocks, on the under sides of stones, on fronds of seaweed, or even within the broken 'bulbs' of the familiar Bladderwrack. The male usually mounts guard, and in the case of some of the Sculpins (Cottidae) may actually 'brood' over them, clasping the egg-masses with its pectoral and pelvic fins, the inner surfaces of which are provided with asperities or hooks to enable it to obtain a firmer grip. Aeration of the eggs is another duty falling upon the male and is generally accomplished by fanning the surrounding water with the pectoral fins. The common or Sand goby (*Gobius minutus*) of our own shores makes a more elaborate shelter, the male first seeking a suitable shell, generally a cockle or small scallop, which it turns over so that the concave side is downwards. It then gets underneath it, clearing away the sand with its tail, until a little chamber has been constructed, communicating with the exterior by a single tunnel-like opening. Finally, it covers the whole structure with fine loose sand, and sets off in search of a suitable mate.

Different Types of Nests

The freshwater sun-fishes (Centrarchidae) scoop out a shallow basin like nest, from the bottom of which all pebbles are carefully removed, leaving a layer of fine sand or gravel to which the fertilized eggs adhere. The work is carried out entirely by the male, who remains to guard the eggs until the youngs are hatched.

Males of Cichlids (Cichlidae) also costruct relatively simple nests, either as a scooped-out hollow or as a hollow on top of a mound. The shape and form of the nest is, of course, partly determined by the nature of the bottom but nevertheless often characteristic of the species. In mouth-brooding species the nest serves merely as spawning site, but in non-mouth brooding species the ova are left in the nest and guarded by both parents. More than one nest may be dug and the eggs transported, in the mouth of a parent, from one nest to another.

In Deep Pits

Two species of African Lung-fish (*Protopterus annectans* and *P. aethiopicus*) prepare a simple but often deep pit or hole, usually in swampy place, along river banks, lake shores, or in true swamps. It seems that the male alone is responsible for preparing the nest, but that it may spawn with two or three females. After spawning, the male remains in the nest, aerating the water by it body movements and even stirring the surface by slow slapping movements of its tail. It also protects the young by snapping at any creature that may attempt to enter the nest from above or below. But, despite this assiduous protection against predators, some manage to slip in and catch the young as they surface to breathe. Perhaps the most successful predator is a species of water spider which is able to station itself without disturbing the water and thus arousing the watchful male.

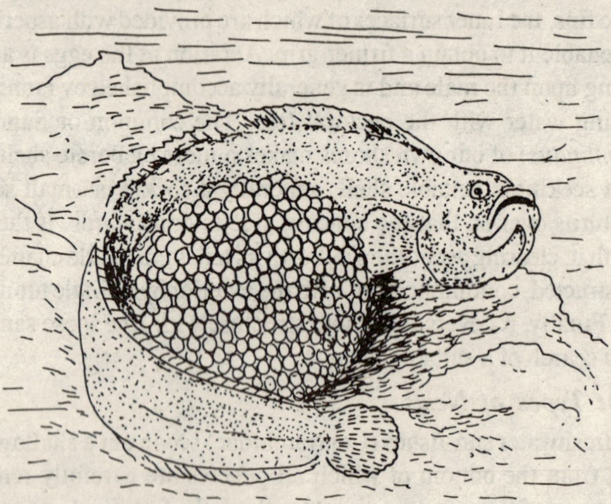

Fig. 17.2. Gunnel or Butterfish (*Pholis gunnellus*) with a mass of spawn.

In closed burrow in Swamp Soil

A third species (*Protopterus dollei*) prepares a more elaborate nest in the form of a closed burrow in swamp soil. The form of this nest rather closely resembles that of the South American Lung fish (*Lepidosiren paradoxa*). Unlike the related African species, the breeding male of the South American form develops highly vascularized filaments on the pelivc fins. The functions of these gill-like structures is still debated. Some suggest that they may serve to secrete oxygen from the fish's blood into the water and thus increase the oxygen conent of the water which is low and rapidly depleted by the developing eggs. Other workers suggest that the filaments act as gills and that the male obtains oxygen from the water, thus reducing the necessity of his making frequent trips to the surface in order to gulp air. Surprisingly, very little research and experiment has been done to elucidate this intriguing question.

In Gelatinous String

The breeding habits of other common British freshwater fishes are varied. Two (Stickleback and Miller's Thumb) are nest-builders. The Pike scatters its numerous eggs amongst aquatic plants but the Perch (*Perca*

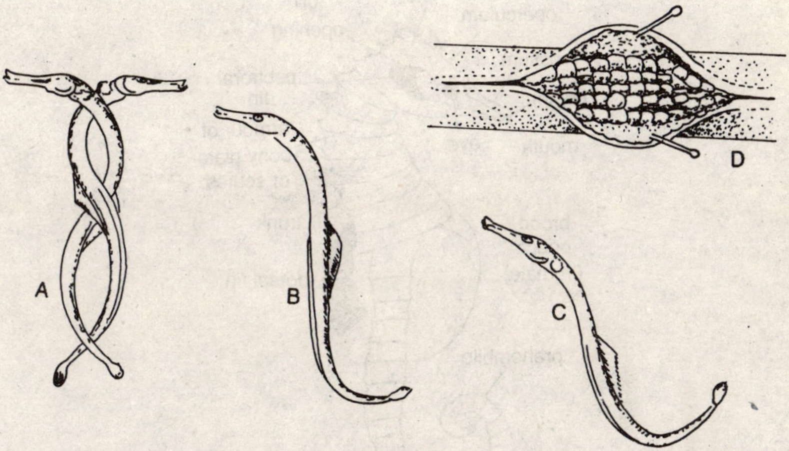

Fig. 17.3. Breeding habits of the Florida Pipe-fish (*Syngnathus floridae*). a—Position of fishes during transfer of eggs; b—Attitude assumed by male while moving eggs backward in the pouch; c—Position of male during period of rest following several egg transfers; d—Portion of pouch opened to show eggs.

fluviatilis) lays its eggs in long gelatinous strings which become attached to the stems of plants or to other submerged objects. The pope or Ruffe (*Acerina cernua*), a relative of the Perch, does not produce egg strings; the eggs adhere singly to sunken branches, plants or stones, Little is known about the breeding biology of the two Loaches (Cobitidoe) occurring in Britain. The Stone Loach (*Nemachilus barbatula*) was once recorded as laying its eggs amongst the submerged roots of an old willow tree.

On Leaves

Finally, we may consider two of the most unusual types of breeding biology, a species that spawns out of water and a group of species which produce embryos capable of surviving in dried-up places. *Copeina arnoldi* (characinoidei) the so-called Splashing Tetra spawns on the underside of leaves overhanging the water surface. The male and female jump together and rest for a moment on the lower surface of the leaf. At this point the adhesive eggs are shed and fertilized. The parents falll back into the water but the male remains near the leaf splashing drops of water on to it with his tail. After two of three days the eggs hatch and the larvae fall into the water. *Copeina arnoldi* is found in the Brazilian Amazon and in Venezuela; it is the only species of the genus with this type of spawning behaviour.

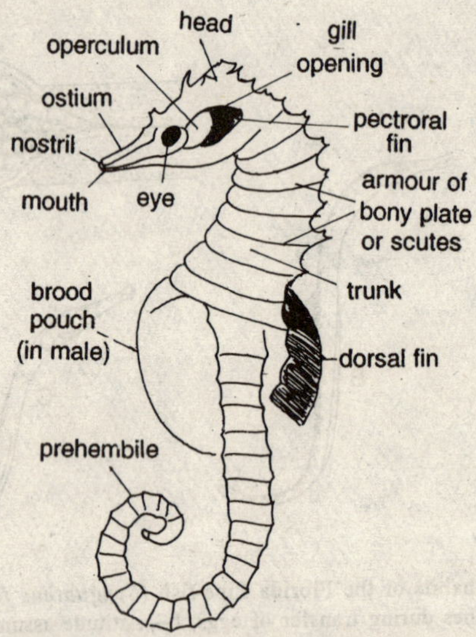

Fig. 17.4. *Hippocampus*.

PARENTAL CARE IN ANNUAL FISHES

In tropical Africa and South America there are a number of small Cyprinodont fishes which produce eggs capable of surviving for some time amongst the bottom detritus of dried-up streams and ponds. These are the 'Annual Fishes' whose adult life is usually confined to a single year. Spawning takes place just before the onset of the dry season. The fertilized eggs are buried in the bottom deposits and embryonic development continues until the water of the stream or pond has evaporated, a rapid process in a tropical dry season. Naturally the parent fishes are killed but the now dormant embryos are protected by the drought resistant egg membrane and are probably protected from complete desiccation by the small quantities of moisture trapped between the particles of soil and humus. Embryonic development is resumed with the return of the waters, the eggs hatching shortly after the floods and the young growing to sapwn and die within that year. Aquarium observations suggest that even if there is no drought the adults die within a year of hatching, the whole biology of the species being geared to an annual life-cycle.

18

Economic Importance of Fishes

Fishes are the rich source of proteins, carbohydrates, fats, vitamins (A, D and E), calcium, iron and many other minerals. Millions of human beings suffer due to malnutrition and hunger and the fishes can form a rich source of fc ɔd and provide a means to tide over food problem of human beings. In addition to solving food problem, fishes provide a number of by-products. Some of the by-products are of high economic value.

Our government is now paying much attention to the development of fish and fisheries for the improvement of quality and quantity of fishes. A number of Fisheries Research Stations have been set up for this purpose. Some of the stations are - Barrackpore in West Bengal, Mandapam in Tamil Nadu, Visakapatnam in Andhra Pradesh etc.

The fishes are useful as well as harmful to us.

USEFUL FISHES

Fishes as human food

The nutritional value of fish meal is higher than the meat of cattles. Fish diet provides fats, vitamins, phosphorous and other minerals. It contains about 50-70% proteins, 12-15% fats and 10-20% minerals (calcium, iodine, phosphorous etc.). All the essential amino acids are present in the proteins. Vitamin A, D, E, B_{12} and other vitamins are also present. Fishes are eaten in various ways such as cooked, raw etc. They deteriorate rapidly after being caught. They must be consumed soon or preserved for the later use.

The principle fresh water fishes consumed as India are :

1.	Major Carps	*Catla catla, Labeo rohita, L. calbasu, Barbus sp., Cirrhina mrigala, C.reba etc.*
2.	Catfishes	*Wallago attu, Mystus seenghala, M.aor, M.cavasius, Callichrous pabda, Silondia silondia, Bagarius bagarius etc.*
3.	Herrings	*Clupea sp., Hilsa hilsa, Gadusia chapra.*

4.	Fether backs	*Notopterus notopterus, N. chitala*
5.	Live fishes	*Clarias batrachus, Heteropneustes fossilis, Channa sp., Anabas testudineus.*
6.	Mullets	*Mugil corsula*
7.	Miscellaneous	*Labeo bata, Tor tor, Puntius, Nandus, Chela, Colisa fasciatus etc.*
8.	Exotic species	*Cyprinus carpio.*

Fish as food of cattles

From both living small and large fishes, the fish meal is prepared. The fishes used for this purpose are sardines, sharks, rays, mackerels and silver bellies. The fishes are chopped and boiled to extract the oil. They are covered with canvas and screw pressed to form cakes. The cakes are dried. The cakes are called *fish meal*. Fish meal contains proteins, fats, minerals, and a number of vitamins. It constitutes a valuable source of food for cattles as well as for poultry.

Name	*Water*	*Protein*	*Fats*	*Food value per pound (calorie)*
Crap	78.9	15.79	4.77	495
Cod	82.6	16.50	0.40	325
Herring	72.5	19.50	7.10	660
Mackerel	73.4	18.70	7.10	645
Perch	78.8	18.45	1.40	402
Salmon	61.4	17.50	17.8	1080
Trout	80.5	17.76	0.74	365

Fish liver oil

The fish oil is the most important by-product of fishery industry. Fish liver oil contains cholesterol and other unsaponifiable substances. It has 55-75% fats, 5-10% proteins, 20-30% water, 0.5-1.32% cholesterol, iodine and vitamins (A, D, E). There is great variations in the fat and vitamin percentage from fish to fish. Cod (*Gadus morrhua*) liver oil is rich in fat contents but poor in vitamin A contents. In some sharks (*Sphyrna blochii, Pristis lupidata etc.*) the fish liver having the highest fat and vitamin contents (about 15000-1000000 lU per gram). The refined liver oil has a medicinal use, being the source of vitamin A and D.

Fish body oil

Besides liver, the oil can also be obtained from the entire body of certain fishes (e.g. salmons, sardines, herrings etc.). The body oil differ from liver oil. For preparing body oil, fish are boiled in large quantity of water. Oil is removed quickly and is washed in boiling salt water. The body oil is poor in vitamin A and D contents. Glycerides of fatty acids are present in varied proportians. Body oil is used for various purposes such as to make paints and varnishes, tanning of skin and dressing of leather, for making cheap soaps, for manufacturing certain chemicals etc. The body oil having large amount of iodine can not be used for cooking purposes. The body oil of sardine fishes contain 19.3% iodine and the salmons about 16%. Fish oil has fungicidal property hence used as fungicides for citrus trees.

Fish manure

Fish that are not fit for human or animal consumption are used to prepare fish manure for the fields. During peak season, the fishes are caught and sun dried by spreading them. The dired fishes are ground in mills and converted to manure. The manure contain 5-7% nitrogen, 4-6% phosphates and 4-6% lime. The manure is very useful for coffee, tea and tobacco crops. Fish *guano* is prepared from the fish material left after extraction of oil. It contains 8-10% nitorogen. Thus the guano is also an effective manure.

Fish glue

The fish glue is prepared from different wastes of fishes such as bones, scales, fins etc. These substances are washed, ground and charged into steam-jacketed cookers. The charge is covered with water and the medium is made acidic with the addition of acetic acid then it is cooked for 6-10 hours. The liquid is extracted from the residue. It is a glue. The glue is used as an adhesive for paper, wood, leather and glass. The glue prepared from the skins of cods is of very high quality and is very useful in photo-engraving work.

Isinglass

Isinglass is obtained from the inner silvery layer of the air bladder of certain fishes. Isinglass is a collagen product. It is insoluble in cold water. Isinglass is very similar to gelatin because it swells in water. It is used in confectionary as a substitute of gelatin. It is also used for the clearification of wines, beer and vinegar. It may also be used for the preparation of plasters and special cement.

Fish leather

The skin of large fishes such as sharks, rays etc. is used for preparing leather. The skin is detached from the fish and soaked in brine for a day. Then the skin is treated with salt solution containing 10% NaCl. The skin is now scraped with a knife in order to remove denticles. They are lined and tanned by usual process. The dried and treated skin is also used for preparing ladies shoes, belts, money bags, suitcases etc.

Fish sausage and ham

These are spiced pasted fish meet prepared and served in Japan, U.S.A. and some countries.

Fish fin

The fin of sharks are exported to China where they are used for preparing soup. They are also smoked sometimes.

Fish Biscuits

There are prepared in Chile and Morocco. Fish flour is belended with biscuit mixture before baking.

For manufacturing of artificial pearl

Scrapping meterial of certain fishes is used as artificial pearls used in jewellery. Scales of garpike are used in jewellery.

For decoration

Several species of fishes are beautifully coloured and are kept in aquaria, ponds and are used for ornamentations. Examples : Gold fishes, angler fishes, globe fishes etc.

Scavangers

Clarias batrachus, Heteropneustes fossilis etc. feed on excreta. Many fishes are omnivourous feeding on decaying matters.

Biological control

Several species of fishes are larvivorous and feed upon the larvae of insects, such as mosquitoes. These are *Punctius, Gambusia, Danio, Rasbora, Channa, Ambassis* etc.

As bait

Several fishes are employed as bait for the catching large fishes and other animals such as turtles.

Insulin

Fish can also provide raw material for the manufacturing of insulin. Shark pancreas is rich in insulin.

Fish and Pharmacopoeia

Certain fishes and their by-products are of great use in Ayurvedic and Unani systems of medicines. For examples sharkes are used for duodenal ulcers, some other fishes for skin diseases. Several fishes are recommended for night blindness, loss of apetite, cold cough, bronchitis, asthama and tuberculosis.

Scientific studies

Some fishes like *Scoliodon*, perch and carps are dissected in laboratories for acquiring knowledge of life.

HARMFUL FISHES

Sharks are extremely dangerous in the sea and injure fishermen and damage their nests. The carnivorous forms destroy valuable food fishes and other useful animals. Certain species of fishes have poisonous spine (sting-ray) and cause painful wounds. Electric fishes (*Torpedo*, *Malapterurus* etc.) are capable of giving electric shock to man. Some fishes (pocupine fish) are poisonous and cause death just in 17 minutes after they are eaten. Large forms such as sword fish damage boats. Some species of fishes are the intermediate hosts of many parasites causing diseases in man and other animals.

19

Origin and Evolution of Paired fins

It is a well developed and universally accepted fact that the earliest fishes possessed no fins but swim entirely by undulations of the body and they were developed later as stabilizing facts to counter act the tendency of body to role side way when in motion.

Two types of fins are developed in fishes: *Unpaired* or *median fins* include *dorsal fin* on middle of the back and *anal fins* along belly behind and *caudal fin* at the hinder end of the fish. *Paired fins* are *pectoral* and *pelvic fins*.

These fins were not present in ancestral forms. Laters are forced to conclude that the paired appendages of vertebrates have no genetic connection with those of invertebrates but have arisen independently as vertebrates novelties but rarely present in ancient *Ostracodermes.*

As many as four theories have been put forward for the origin of the paired fins: -

(i) Gill arch theory

(ii) External gill theory

(iii) Lateral fins fold theory

(iv) Ostracoderm theory

(i) Gillarch Theory

This theory is put forward by *Gagenbaur.* According to him the paired fins with their girdles are derived from the gill arches. The two pairs of gill arches transformed into the pectoral and pelvic girdles and their gill rays gave rise to skeleton of the fins. Actually this theory was proposed on assumption that the primitive gnathostomes possessed gill arches extending further behind than they do at the present time. Soon on the basis of this theory the direction of fins has evidently been dorso-ventral and both the pectoral and pelvic fins were very close together behind the head. But the present position of pelvic fins is explained by their migration to say backwards.

Objections. This theory objected by a large number of considerations.-

1. It offers no explanation of the participation of a large number of muscle segments, in the formation of fins.
2. In ontogeny, paired fins never take appearance as a dorsoventral fold. On the contrary it always arises as a more or less extended longitudinal ridges.
3. The position of pelvic fins was attributed by *Gagenbaur* to their backward migration but neither in the primitive fishes nor in the early fossils. This is observed.
4. Presence of vertical muscle buds in front of pelvic fin fold in ontogeny has been supposed to indicate backward migration. Since such buds are also found behind same fins. This can hardly be so.
5. In the skeleton, fins were derived from the gill rays, the muscle supply should come from the lateral gill plates but not from the myotomes.
6. The position of limb-girdles is lined out side the blood vessels and nerves and coelom is the reverse of what it should be if derived from gill arches.
7. The theory does not explain the remarkable resemblance borne by the paired fins to the unpaired fins. So all the objections impart deadly against to the gill arch theory.

(ii) External Gill Theory

It is variant of gill arch theory and put forward by *Grahm kerr*, according to which the paired fins have been derived from the external gills which occur in crossopterygians, lung fishes and urodele amphibians.

It is inferred that probably external gills are present in each gill arch. It is also stated that gill arches extended tailwards beyond the branchial limit of existing vertebrates. Such external gill arches are considered as potential organs of support and movement as shown by their form and musculature in urodeles.

Besides the external gills occur at a certain time and for special purpose of only respiration. Except for a superficial resemblance to the highly specialized and reduced limbs of "*Dipneumnous*" external gills differ from paired fins in every important respect.

(iii) Lateral fin fold theory

The modern version of this theory was put forward by *Balfour, Tracher* and *Mioast.* This theory holds that the paired fins are essentially of the same nature as the unpaired fins and have been derived from longitudinal folds. Their apicals have been arisen by the stiffening of the

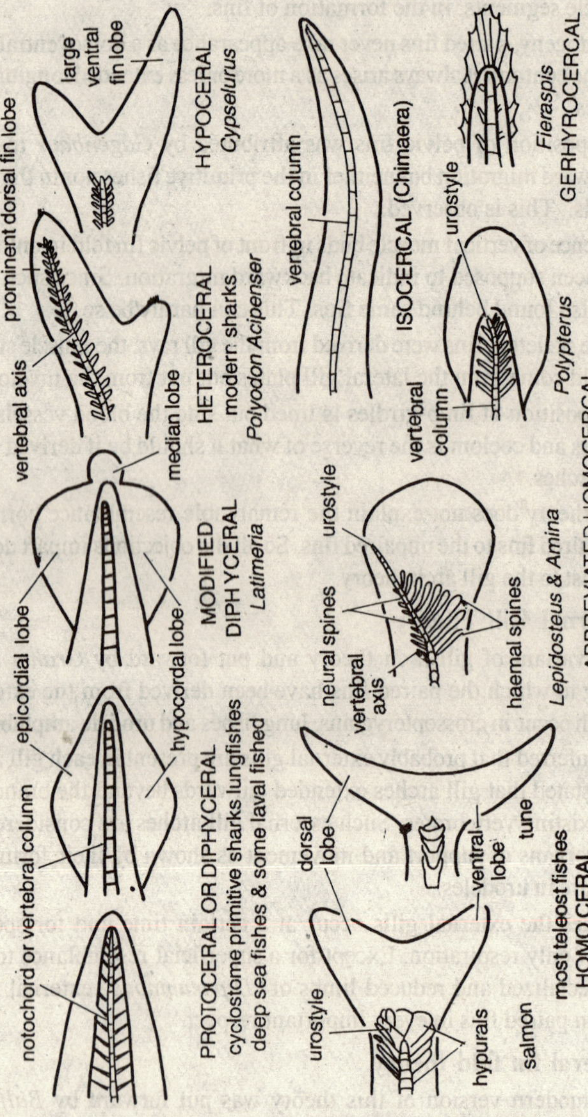

Fig. 19.1. Different types of Caudal fins.

fin fold and girdle by extension inwards of the basis of radials. The continuity of the pectoral fin with the pelvic fin is not an essential part of the theory. But the important thing is the recognition that the paired fins arose as longitudinal fins similar to those which had given rise to the median fins. Possibly the paired fins, were discontinous and differentiated into the pectoral and pelvic fins from the very beginning.

There are many facts in support of the continuity of pectoral and pelvic fins, like great extensions of muscle band areas and of collector nerves both in front and behind the fins. The concentration of fin elements and shortening of fin rays and the frequent presence of the great number of muscle bands in early stages than incorporated in the fins.

Very remarkable in this connection is the production of an uninterrupted series of muscle bands from the pectoral to the pelvic fin in some fishes such as *Scyllium* and *Lophies*.

Every trunk segments of the body is capable of producing paired fin elements muscular, nervous and skeletal. The potentiality is, however, expressed by segments along restricted areas for instance in *Scyllium* 2 to 19 segments contribute to pectoral fin and 25 to 35 pelvic fins. The motor components of the special nerves always faithful to its myotomes through out the changes which may take place in the ontogeny. The number of spinal nerves sending motor fibers to the nerve plexus of the fin indicates the number of muscle segments that are incorporated in the fin.

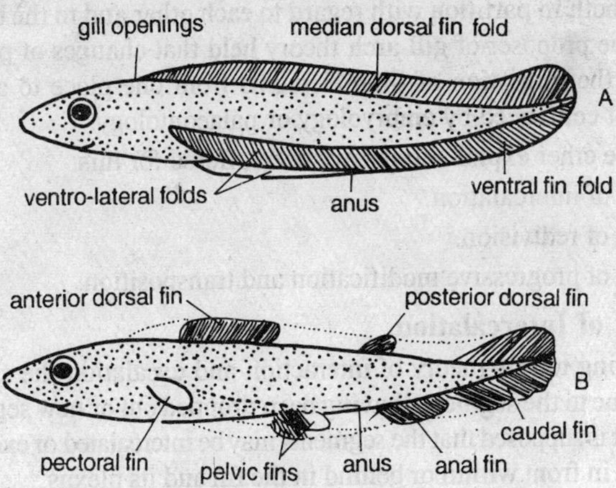

Fig. 19.2. Organ of fins Balfour's Fin Fold Theory.

Owing to the above account this theory is widely accepted and has its confirmation from palaeontology and embryology.

(iv) The Ostracoderm Theory

According to this idea that the appearance of paired appendages goes to the ostracoderms which may have been the remote ostracoderms had a cephalothorax with a shield of horny plates. Certain forms possessed lateral fleshy lobes projecting from lateral body wall. These fleshy lobes may have become pectoral fins and the pectoral girdle may have developed as ingrowth from the basal portion of these fins. The dermal bones of the pectoral girdle may have originated from the bony plates of the thoracic shields.

Another group of Ostracoderms had a paired row of dermal spines on the ventral side of the body, not unlike the spines of Acanthodian sharks. The spines may have been in use in clinging to a rapidly moving fresh water streams. Loss of and relation of spines except for those in the pectoral and pelvic regions might account for the origin of paired appendages from the ostracoderm ancestor with the growth of the muscles from the myotomes into the bases of these spiny fins and their movement may have been accomplished.

The anatomical, embryological and fossil evidences show that the fin fold theory is inadequate because it arose from the median and paired spines of ostracoderm.

The paired fins frequently changed their position of body in the course of phylogeny. This variation is even found in individuals of same species and effects both in partition with regard to each other and to the body as a whole. The proposer of gill arch theory held that changes of position and due to the migration of limb elements from one place to another which is not confirmed by embryology or palaeontology.

So three other explanations have been offered for this:

(i) Theory of intercalation.

(ii) Theory of redivision.

(iii) Theory of progressive modification and transposition.

(i) Theory of Intercalation

According to this theory of interaction and excalation the shifting happened due to the segments having been dropped out or new segments adding up. It is supposed that the segments may be intercalated or excalated at any point in front within or behind in the fin and its plexus.

If such was the case there should be joined of growth and joines of reduction, their new segments appear and old vanish. But no such segments occur and 2 sets of fins, the median and paired shift independently. But the first dorsal is opposite to pectoral in lamina between the pectoral and pelvic, in *Alopecies* opposite pelvic in *Scyllium* and behind pelvic in *Raja*. If the dorsal fin is homologous in four genera no addition and no suppression of segments can account for this disposition. This and other objections have been successfully adequated against this theory.

(ii) Theory of Redivision

This theory states that one individual organ is composed of 20 segments and other 19 segments. The difference according to this theory is not due to addition or substraction of a segment but due to the subdivision of the individual organ into 20,21 or 19 segments. On the bases of this theory no segments in the first case strictly corresponds to any segment in the other two. This theory offers no explanation and cannot be applied in any way better than the theory of intercalation.

(iii) Theory of Progressive migration and Transposition

According to this theory by progressive growth in one direction and by corresponding reduction in the other, change of position takes place. By gradual growth a limb supplied with nerves may come. A limb plexus may be shifted without alternating its structure from one region to another by growth at one end accompanied by reduction at the other. The explanation is incomplete hormony with the findings of comparative anatomy and morphology.

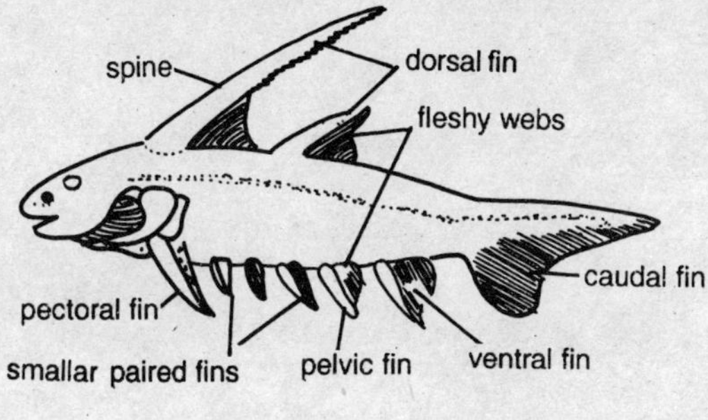

Fig. 19.3. Fin spine Theory.

Conclusion

By all the above theories we are bound to conclude that the lateral fold theory alone agrees with the facts provided by the comparative anatomy and embryology.

As has already been shown palaeontology affords some evidences to the existence of paired fin folds. The absence of any paired fins in cyclostome, has always been a differently in adopting the view that the ancestral craniates were provided with paired folds. But if we adapt such conclusion that the modern cyclostomes are the degenerate descendent of ancient ostracodermes, it seems probable that Lampreys and hag fishes have lost the paired fin folds, for indication of such folds has been found in more ostracoderms.

Whatever may be the case it is definite that paired fins are originated from the lateral fin folds, even which the ancestral form is represented the original picture is immaterial and its evolution firstly took place among the fishes as a result of physological rearrangement.

20

Origin and Evolution of Fishes

Although the word fish is often used to include any animal living in water. The fish may actually be defined as a vertebrate living a purely aquatic life, propelling and buoyancing it by fins and breathing by gills.

The Ist chordate known as a fossil is *Jamoytius*. The remains are very fragmentary and give no clue as to what the animal represents by them look like.

In a broad sense the Ist vertebrate were fishes. Actually the Ist fishes were primitive jawless forms known as Agnatha (*A*=without, *gnathas*=jaws). From these early aganth vertebrates the Gnathostomata or jawed vertebrates evolved. The Agnatha falls under two groups the *Ostracodemi* and *Cyclostomata*.

Turning to the fossil record Agantha was found in the *Ordovician* and *Silurian* deposits and *Devonian* period had variously diverged types. The best known examples were the *Cephalaspis* in the general pattern of a fish. The head was covered with a head shield and the trunk and tail were very much like those of a modern fish. On either side of the body were paired appendages. The head with shields and the body had the dorsal and caudal fins. The anterior part of the trunk was covered with a transverse band like scales which pass into more typical scales to the tail.

The endoskeleton was completely ossified as in modern fishes. A striking peculiarity was of the electric organs as in modern electric fishes.

Other ostracoderms like *Anaspids* were small headed fish-like forms with plates on the top of head. It seems not impalpable that gnathostomes were derived from this Anapsid stem. At the close of Devonian period the ostracoderm disappeared posing the battle with the Gnathostomata. Although the Silurian and Devonian Agnatha were ancestral to Gnathostomata, the fossils fail to show such a link. It is because the Silurian and Devonian ostracoderms were specialised and approaching the end of several evolutionary lines.

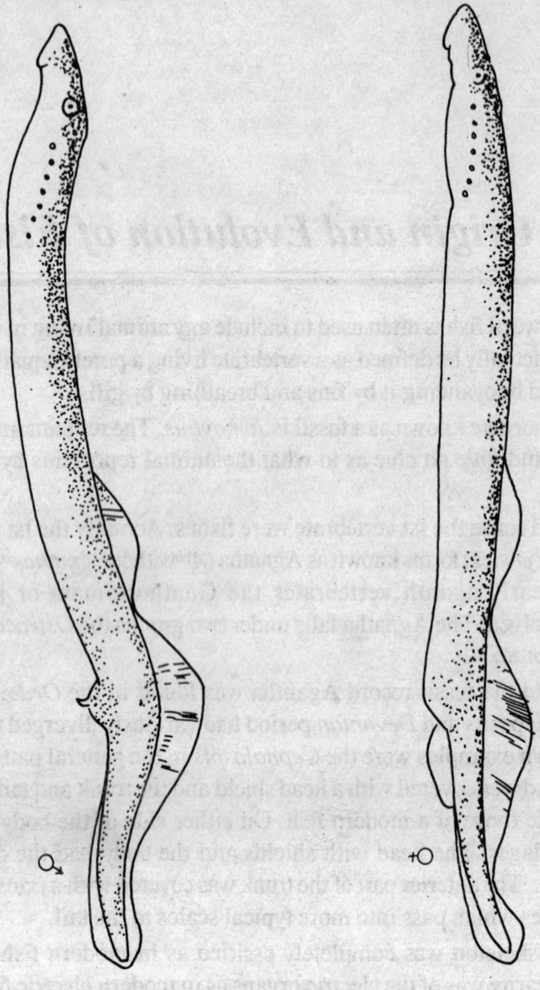

Fig. 20.1. Lamprey.

Cyclostomes are swimming jawless, limbless, vertebrates falling in two groups the lampreys and hagfishes, which are products of two lines of evolution. Lampreys are degenerate offshoot of *Cephalospidomorpha*, while hagfishes are degenerate decendant of *Pleurospidomorpha.*

One of the greatest step in the evolutionary history of vertebrates was the appearance of jaws, as it opened new lines of adaptation beyond bottom feeding or parasitic life. The jaws arose from the gill arches. Ostracoderms had a large number of gill arches which were as many as 10, in *Cephalaspis*, possibly two of these gill arches were eliminated and the remaining were transformed into jaws. The jawed vertebrates were the upper Silurian and lower Devonian forms included in the class Placodermi. These ancient gnathostomes evolved and burrid during early and,middle Palaeozoic time and for sometime dominated life in water. The dominance, however, was short because by the end of Devonian period most of them disappeared and by the end of *Palaeozic era* all of the placoderms had vanished.

The placoderms were, however, a heterogenous lot and taken together may not be a natural group. Placodermes may be regarded only as an ancient experiment in the evolution of jawed vertebrates. In.many respect they were specialized and no group from them can be looked upon as an ancestor of the higher fishes. It is, however, evident that placoderms evolved from primitive gnathostomes and when taken together they indicate many characters expected in the ancestral jawed vertebrates. It is, therefore, possible to realise the Ist step in the evolution beyond the jawless condition by projecting back from the known placoderms.

The rise and multiplication of sharks and bony fishes contributed to the disappearance of ostracoderms and placoderms. The factors responsible were the stream lining of the body by development of median and paired fins with rays, rhythmic and alternate passing down of muscular rays along side of body, specialization of hyoid arch for joining the jaws with the cranium. Both the groups appeared in Devonian period more possibly in the Silurian period. The shark evolved rapidly in the Devonian time and continued their expansion upto the Carboniferous and Permain periods. At the close of Palacozoic era many lines of shark evolution out but in restricted numbers.

Sharks are generally considered to be primitive fishes although they appeared in the fossil record at a stage later than at which the bony fishes appeared. Perhaps the allocation of sharks to a primitive position is grown out of the view that the cartilaginous skeleton is more primitive than the bony skeleton.

Yet it is equally reasonable to think that the cartilaginous skeleton of shark is secondary and the bony skeleton of ostracoderms is truely primitive. Certain other characters typical of sharks as group are internal fertilization, no air bladder, seperate gill openings and marine habitate.

One of the shark known from fossil evidences was *Cladoselache* found from the upper Devonian deposits. There was large heterocercal tail in which the two lobes were outwardly equal. There were 2 dorsal fins, pectoral and pelvic fins and a pair of fins on either side of the base of tail.

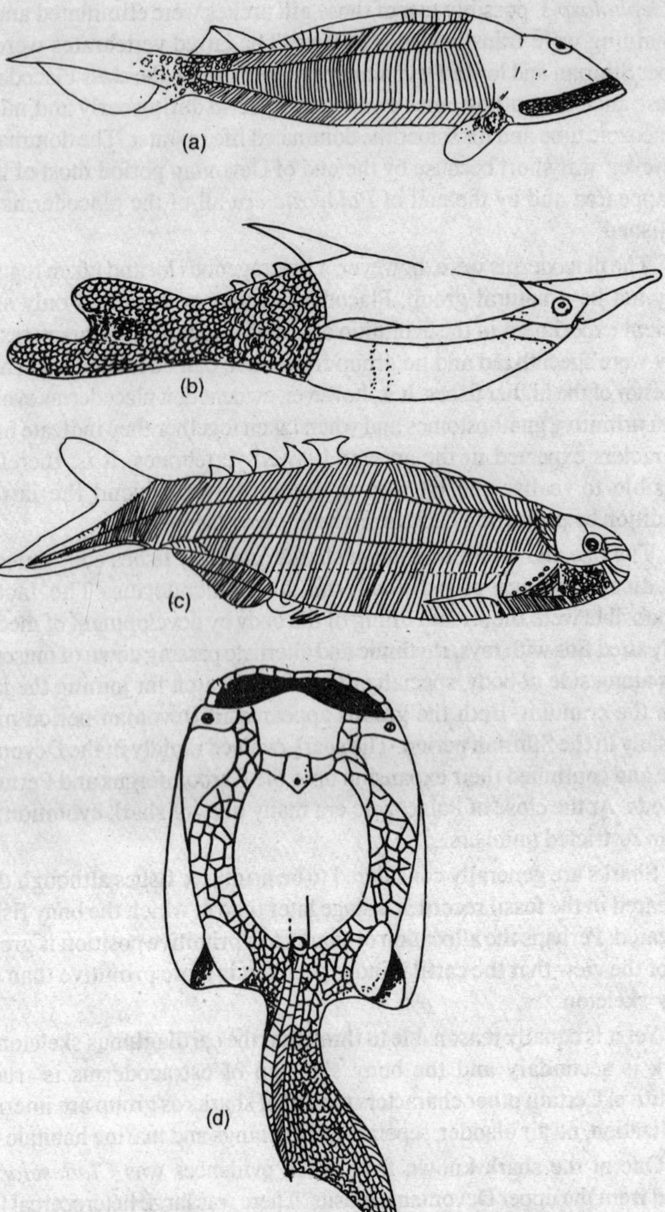

Fig. 20.2. Representatives of Ostracoderms.

Eyes were large and were facing forward on the skull. A jaw suspensorium was amphistylic which is primitive condition for jawed fishes. Each tooth has a high central cusp and two lateral cusps is primitive and found in ancient fossil forms. The structural pattern of *Clado-selache* is primitive for shark and fish may be consider to approximate to the central structure from which later sharks evolved in high direction. (Pleurocanth sharks, typical sharks, skates and rays, Bradiodonts and Chimaeroid). As a result of these evolutionary changes the cladoselachians were gradually replaced by these forms that sprang from them. Yet the cladoselachians were so well established that they continued up to the end of *Palaeozoic era*.

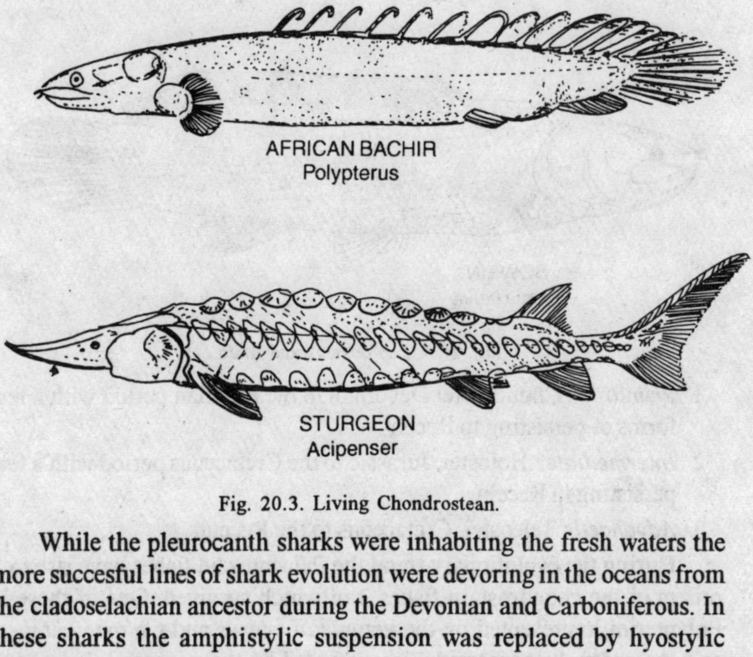

Fig. 20.3. Living Chondrostean.

While the pleurocanth sharks were inhabiting the fresh waters the more succesful lines of shark evolution were devoring in the oceans from the cladoselachian ancestor during the Devonian and Carboniferous. In these sharks the amphistylic suspension was replaced by hyostylic suspension. There exists even today certain sharks that are connecting links between the modern sharks and cladoselachians. These are *Heterodontus* and *Chlamydoselaches*, with amphistylic type of jaw suspensorium.

The other group of shark-like fishes are the Brychiodonts known with crushing type of deviation and the Chimaeroids which persisted at the ancient time in deep oceanic water.

Of all the animals that lived in water none has been so successful as the bony fishes which invaded all types of waters i.e. from streams to oceans. Bony fishes are of ancient history. They Ist appeared in the fresh water deposite of mid Devonian time. The Ist bony fishes belong to group Palaeoniscoidea examplified by the Devonian genus *Cheirolepis*. From an ancestory illustrated by *Cheirolepis* the main line of bony fishes the Actionpterygian was evolved passing through 3 stages.

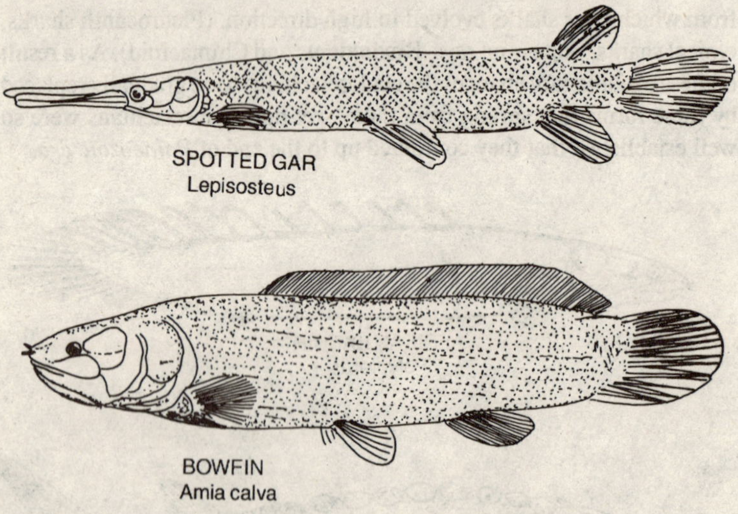

Fig. 20.4. *Lepidosteus* and *Amia*.

1. *Primitive* : Chondrostei Devonian to the Permian period with a few forms of persisting to Recent.
2. *Intermediate* : Holostei, Jurassic to the Cretaceous period with a few persisting in Recent.
3. *Advanced* : Teleostei, Cretaceous to the Recent.

During the evolutionary trend the Palaeoniscid fishes gave rise to 3 orders of the chondrostain fishes 2 of which survived. One of these is Polypterini, represented by the genus *Polypterus* and *Calamoichthyes*, and the order Acipenseroidei, represented by the genera *Polydon* and *Acipenser*.

During the Traissic period another order, the Holostei, sprang from the Palaeoniscoid ancestors. These fishes were intermediates between chond-rostein and teleostean.

As the chondrostein fishes declined at the end of Triassic period they were replaced by holostein fishes from the subholostein stem. The Ist holostein fish is represented by *Semionotus*. The culimination of holostein evolution was reached in the Jurassic and early Cretaceous time. After which there was a decline. However, two genea of these fishes have persisted to the recent time *Lepidosteus* and *Amia*.

With the onset of Cretaceous period the holosteins were replaced by Teleost fishes. The Ist teleost was represented by genus *Leptolepis* appearing in the Jurassic period which was intermediate between holostein and teleostin fishes. By the Cretaceous period the Teleost fishes were well established and following numerous lines of adaptive radiation established as Masters of all types of waters.

A simplified arrangement of this radiation may be represented by the orders:

1. *Isospondyli* : The Ist and least specialised teleost fishes examplified by salmon and trout.
2. *Ostariophysi* : A group predominantly fresh water with a chain of Weberian ossicles connecting the ear with the air bladder e.g. carps and cat-fishes.
3. *Apodes* : Elongated fishes in which the paired fins are frequently supressed e.g. True-eels.
4. *Heteromi* : Deep sea fishes with long weeds and with the caudal fin supressed.
5. *Malacichthyes* : Intermediate teleosts with various progressive features e.g. Sea-horse, pipe-fishes.
6. *Acanthopterygii* : The most advanced and most numerous of teleost fishes representing the culmination of fish evolution eg. *Ophiocephalus*.

In the Devonian time another group of fishes, the choanate fishes, evolved in the begining. The Ist choanate fish, was similar to Ist Actinopterygian fish *Cheirolepis*. The *Dipterus* and *Osteolepis*

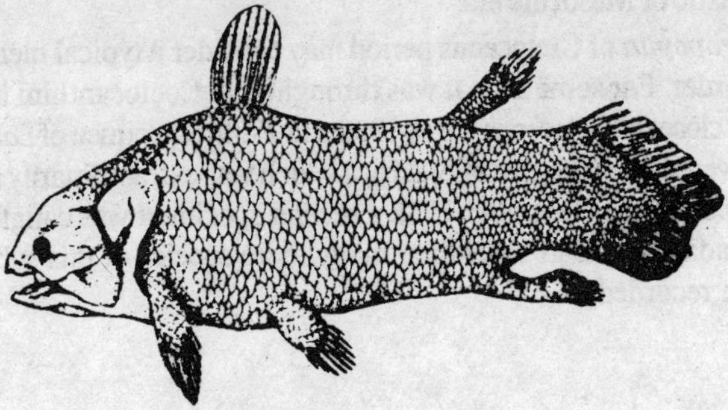

Fig. 20.5. *Latimeria*.

representing the ancestral types of the two lines of chordate evolution. The Dipnoan and Crossopterygians were, however, significantly different from *Cheirolepis*.

The lung fishes are represented by three genera, at the Ist time *Neoceratodus* in Australia, *Protopterus* in Africa and *Lepidosiren* in S.America. The ability of these lung fishes to breath air is suggestive of an intermediate stage between fishes and land vertebrates. The total evidences, however, indicate that the fishes have never been on the direct line of evolution leading from fishes to lst living vertebrates. The earliest long fishes are represented by genus *Dipetrus* of the Devonian time. From *Dipterus* the long fishes evolved the central line of evolution lead the *Ceratodus* of the Triassic period with *Cepiceratoclus*, as its direct descendants. *Protopterus* and *Lepidosiren* have evolved as side branches from the central stem of Dipnoan evolution.

The Crossopterygian fishes like Dipnoans appear in the middle Devonian time. They possessed certain chordate characters as shown by the Devonian genus *Osteolepis*. From the basal stock of the crossopterygian evolved one sub-order of Rhipidistia and the other sub-order Coelacanthini. *Osteolepis* was a member of Rhipidistia. Rhipidistians were primarily fishes which began in the Devonian time as *Osteolepis* and followed a dichotomus evolution. The remaining of the Palaeozoic era after which they become extinct. One line was that of Holodonts characterised by *Holotypias* of Devonian time which showed an evolution parallel with lung fishes. The other line was Rhizodonts of which the Devonian Genus *Eusthenopteron* is of particular importance as these fishes on the direct line of evolution towards amphbians.

The other groups is Crossopterygian, Coelocanthini are removed from the main line of evolution towards vertebrates. They were marine and characteristic of Mesozoic era.

Macropomia of Cretaceous period may consider a typical member of this sub-order. For some time it was throught that Coelocanthini became exitinet at close of Cretaceous but in 1938 and 1939 a specimen of *Latimeria* was discovered from the east of S. Africa which was extraordinarily similar to *Macropomia*. In December 1952 a second specimen was caught from cost of Madagascar and since then many more species of this sub-order have been recorded.

Index